THE ESSENCE OF

DISCRETE MATHEMATICS

THE ESSENCE OF COMPUTING SERIES

Forthcoming titles:

THE ESSENCE OF

DISCRETE MATHEMATICS

Neville Dean

An imprint of **Pearson Education**

Harlow, England · London · New York · Reading, Massachusetts · San Francisco
Toronto · Don Mills, Ontario · Sydney · Tokyo · Singapore · Hong Kong · Seoul
Taipei · Cape Town · Madrid · Mexico City · Amsterdam · Munich · Paris · Milan

Pearson Education Limited
Edinburgh Gate
Harlow
Essex, CM20 2JE
England

and Associated Companies throughout the world

Visit us on the World Wide Web at:
http://www.pearsoned.co.uk

Typeset from the author's disks by Dorwyn Ltd, Rowlands Castle, Hants

Printed and bound in Great Britain by
TJ International Ltd, Padstow, Cornwall

Library of Congress Cataloging-in-Publication Data

Dean, Neville.
  The essence of discrete mathematics / Neville Dean.
  p.   cm.
  Includes index.
  ISBN 0-13-345943-8
  1. Mathematics.   2 . Computer science—Mathematics.   I. Title.
QA39. 2 . D43 1996
511– d c20   96-8580
                                                                    CIP

British Library Cataloguing in Publication Data

A catalogue record for this book is available from
the British Library

ISBN 0-13-345943-8

---

10 9 8 7 6
07 06 05 04 03

Contents

Series Preface

As the consulting editor for the Essence of Computing Series it is my role to encourage the production of well-focused, high-quality textbooks at prices which students can afford. Since most computing courses are modular in structure, we aim to produce books which will cover the essential material for a typical module.

I want to maintain a consistent style for the Series so that whenever you pick up an Essence book you know what to expect. For example, each book contains important features such as end of chapter summaries and exercises, and a glossary of terms, if appropriate. Of course, the quality of the Series depends crucially on the skills of its authors and all the books are written by lecturers who have honed their material in the classroom. Each book in the Series takes a pragmatic approach and emphasises practical examples and case studies.

Our aim is that each book will become essential reading material for students attending core modules in Computing. However, we expect students to want to go beyond the Essence books and so all books contain guidance on further reading and related work.

In recent years there has been increasing emphasis on formalism in computing. *The Essence of Discrete Mathematics* aims to provide the fundamental mathematical skills to be able to cope with the formal aspects of computing. This book provides a straightforward mathematical foundation for those students who currently lack the mathematical maturity to pursue more advanced topics such as algorithmics and formal methods in software engineering. This is a good example of a book written by a practising teacher who could not find a suitable textbook for his course, so he wrote one!

Computing is constantly evolving and so the teaching of the subject also has to change. Therefore the Series has to be dynamic, responding to new trends in computing and extending into new areas of interest. We need feedback from our readers to guide us – are we hitting the target? Are there 'hot' topics which we have not covered yet? Feedback is always welcome but most of all I hope you find this book useful!

Ray Welland
Department of Computing Science
University of Glasgow
(e-mail: ray@dcs.gla.ac.uk)

Preface

This book has arisen out of an introductory course in discrete mathematics given over five years at Anglia Polytechnic University. Like many other colleagues teaching discrete mathematics around the world, I quickly began to realize that it was not an easy or popular subject with the students. A major factor in this difficulty was the students' poor understanding of sets and logic; all the textbooks seemed to pass quickly over these basics (if at all) before plunging headlong into the traditional topics of discrete mathematics, such as proof, graph theory and combinatorics. It was my contention that these more advanced topics would 'take care of themselves' if only more time were devoted to ensuring a firm foundation — subsequent experience seems to confirm that contention. Owing to a lack of suitable texts, I found it necessary to write a book about the *essence* of discrete mathematics; this is the result of that effort.

My approach in *The Essence of Discrete Mathematics* is to help the reader grow in mathematical maturity, and in particular towards an understanding of the basic concepts of discrete mathematics. Initially this understading will be very much incomplete and flawed (or 'wrong' in mathematical parlance!). In my experience, students are only too happy to accept their knowledge is lacking *provided* their teachers recognize and accept this; under such circumstances, students can and will learn to revise and adapt their understanding as part of a continuing growth. When I have attempted to explain to students that their understanding of a particular concept is not strictly correct, the usual response is 'Yes! I hear what you're saying, but for the moment I need a simplified view'.

Following an introductory chapter, which attempts to explain to the reader how to use the book to best effect, there follow chapters on sets and logic. In these chapters, an effort has been made to give the reader clear instructions on how to 'calculate' values for mathematical expressions for small finite sets. Since it is best to use examples familiar to the reader, sets of numbers are used extensively. Nevertheless, it is important for the reader to see non-numerical examples, and so a case study is introduced at the end of the chapter on sets; this case study is subsequently used throughout the remainder of the book.

Once the ideas and some basic skills of handling sets and logic have been mastered, the reader can progress to chapters on relations and functions. Relations and functions are first introduced as intuitive notions before it is explained how they can be modelled using sets; as is conventional, the distinction between a relation or function and its model as a set of ordered pairs is soon dropped for sake of brevity and clarity.

There follows a chapter showing how the ideas of modelling with sets and logic can be applied to more practical problems. Finally there is a brief concluding chapter which invites all readers to continue their mathematical growth; *The Essence of Discrete Mathematics* is meant to be the beginning and not the end.

Acknowledgements

Many people have helped at all stages in the production of this book. I would like to thank my colleagues at Anglia Polytechnic University for their general support and comments; my students who have given very useful feedback on the book, and who have done a great job spotting errors in the answers to the exercises; Mike Hinchey of the New Jersey Institute of Technology and the University of Limerick for encouraging me to write the book in the first place, and for many fruitful discussions over a pint of Guinness; and Jackie Harbor of Prentice Hall for her patience.

The book has been produced using $\LaTeX 2_\varepsilon$ (principally emTeXand gTeX) on various PCs. I am indebted to Anglia Polytechnic University for the use of facilities in writing this book.

It is right that authors express deep gratitude to mothers and fathers, partners, children, teachers and close friends; no less than any other author, I too would like to express such heartfelt gratitude to all who have loved and supported me, and who have helped me over the years. But I am especially thankful to my late father, Wilfred, who was a wonderful Dad and a great teacher. If there is anything of value in this book, then it is due in no small measure to him.

To the memory of my father,
without whom so much
would have been so little.

CHAPTER 1

READ.ME

1.1   The nature and purpose of discrete mathematics

Mathematics exists to make life easier. To many people this may seem a rather odd statement, especially those who struggled with the subject at school and for whom there seems no application of mathematics to real problems. Part of this difficulty may lie in a lack of understanding of the nature of mathematics and the way in which it can be used. In particular, mathematics and its application requires a different way of thinking and reasoning to what most people do naturally (for want of a better term I shall refer to this natural way of thinking as 'verbal reasoning'). Fortunately most, though not all, human activities do not require a mathematical approach.

Yet there are many activities which either need or could benefit greatly from the application of mathematics. The field of informatics is one such area. The particular branch of mathematics which is needed is called *discrete mathematics*. The purpose of this book is to present the underlying foundations of discrete mathematics, to develop the ability of the reader to think in a more mathematical way, to show how this mathematics can be applied and to encourage the reader to apply these skills and knowledge.

Before attempting to understand the rôle of mathematics (how it can be applied and how it can be learnt) it is useful to consider the concept of a *model*.

1.1.1   Modelling

The notion of a model is central to most applications of mathematics and to computing. A model can be thought of as being built from model objects which correspond to real world objects. For example a wooden model of an ocean liner can be built up from pieces of wood to correspond to the various decks, funnels, lifeboats, hull, keel, propeller screw and so on. Now an essential feature of a model is that it shares certain properties with the system it models; we say that the model *captures* properties of the real system. Thus in our liner model we might label various sides of the piece of wood which models the hull as 'top', 'bottom', 'bow' and 'stern'.

These sides correspond to the top, bottom, bow and stern on the real ship. Note how it is strictly necessary to talk about 'the piece of wood which models the hull'. This is a rather awkward way of talking about the model, and so we would

1

normally say simply 'the hull' where it is clear from the context that we are talking about the model rather than the real hull. Similarly a phrase such as 'the side of the piece of wood modelling the hull which is labelled "bow" ' will be usually referred to simply as 'the bow'; clearly there are enormous advantages in this convention. Now in our model, we could place the funnels at the top of the hull, the keel along the bottom of the hull and the propeller screw on the bottom of the stern. Our model has captured the *relations* between the component parts of the real liner. Much of discrete mathematics is about capturing just such relations, except that we shall use mathematical objects known as *sets* rather than pieces of wood.

For the purpose of capturing the relations between component parts, the model can be any shape we like, possibly built out of box shaped pieces of wood. If we want to capture more information, however, we might decide to make a scale model, at least of the main features. Suppose we have a 1:100 scale model. A distance of 1cm in the model corresponds to a distance of 1m in the real world system so that sizes of objects in the model correspond to those of the objects they represent on the liner. Furthermore the shape of the model is the same as that of the real liner except that all dimensions are 100 times smaller. Again it is possible to use mathematical formulae to model shapes of surfaces and dimensions rather than pieces of wood; this sort of mathematics (sometimes called continuum or continuous mathematics) is the kind that most people have already met at school, and often found difficult. You may be pleased to learn that this type of mathematics does not form part of discrete mathematics and will not occur in this book.

Of course, the liner will still have properties which are not captured by our scale model; for example, the materials used in the construction may not be represented in the model. (Ocean liners are usually not all wooden these days!) It might be tempting to modify our model by replacing the wood by appropriate materials. This would not necessarily be the most convenient way of capturing the information however. Instead we could list the various components and the materials from which they are to be made; a verbal model in fact. A similar principle applies when using mathematical models; the mathematical model should not necessarily try to capture all aspects of the system. Sadly this is all too often overlooked.

Yet another type of model could consist of scale drawings of cross sections of the liner, perhaps augmented with mathematical formulae to represent the shapes of curves at various points.

1.1.2 Using models and the nature of learning

Models capture properties about a real world object or system, whether by using pieces of wood, pictures, words or mathematical notation. As such they can be used in two different ways:

- Models can be used in constructing real world objects; for example, the mathematical equations used to describe the shape of the liner can be used to ensure that the real ship has the right shape.

- Models can be used to understand objects, processes and concepts; for example, someone who is familiar with a scale model of the liner QEII should have no difficulty in recognizing the ship in reality.

Much of learning is based upon constructing in our minds models of new ideas and techniques which are based upon existing experiences and ideas. For example, word processors are usually so designed that beginners can think of them in terms of typing words onto paper: a mental model is built in which the screen display corresponds to a sheet of paper with words typed on it. Of course such a model has its limitations; for example the beginner must soon learn that it is not necessary nor desirable to press the return key at the end of each line, and that pressing the return key is used to indicate the end of a paragraph. Needless to say, this *refinement* of the model requires an understanding of what a paragraph is. Gradually the mental model is developed so that the operator becomes more proficient. But this is only achieved by getting a lot of experience and by making the effort to incorporate this experience into the model. Naturally some guidance and explanation from a book or a tutor will help.

Similar principles apply to learning concepts in computing whether in a general area (artificial intelligence, say) or the commands and constructs of a specific language (such as C++). Now it happens that many of these concepts can be modelled using the essentials of discrete mathematics that are covered in this book; if you understand this mathematics, then learning new ideas in computing can be made much, much easier. (Computing lecturers and authors please take note!) Thus there are two reasons for learning discrete mathematics:

- it is very useful to help us build systems;
- it can make learning so much easier.

1.2   Learning discrete mathematics

The principles of model building as a way of learning apply also to learning discrete mathematics itself. In fact 'discrete mathematics' is a very wide subject area, and textbooks exist which are over a thousand pages long but which are still introductory. But most of the content of discrete mathematics is based upon a few mathematical ideas - the 'Essence of Discrete Mathematics'. These basic ideas are sets and logic. The purpose of this book is to help you learn and understand this essence.

Now in order for you to build mental models of the essential concepts of discrete mathematics, the following are required:

- some basic concepts with which you are already familiar – these are set out later in this section;
- lots of experience – this you must obtain by working through the examples and exercises in this book;

- the incorporation of this experience into your model – this requires you to observe and think carefully about what you do in the examples and exercises (rather than just concentrating on getting the right answers).

Needless to say this learning process is not easy, nor can it be done in a hurry. Nevertheless it is very worthwhile, and you should find working through the book an enjoyable challenge.

1.2.1   The building blocks for learning discrete mathematics

The essential concepts of discrete mathematics described in this book are built up from the general notion of a procedure: a procedure takes an input and produces some sort of output. This approach has been adopted not only because it is a concept with which most readers should be familiar, but also because it helps relate the mathematics more closely to the ideas of computing.

The mathematical ideas of sets and logic are first developed, and then these ideas are used to model functions and relations.

Functions can be regarded as procedures. For example, giving the number 2 as input to the 'square function' gives the number 4 as output, the function having carried out some process on the input value. Most programming languages have some means for defining functions; some languages, known as functional languages, are based entirely on the notion of functions.

Another basic idea which can be thought of in terms of procedures is that of a relation. The idea of a relation has already been introduced in the discussion of modelling a liner; the relation between funnel and hull is that the hull is underneath the funnels. Relations are fundamental to much of human understanding and communication, which is why small children are encouraged to build models of ships out of empty boxes. In this book, you will meet several ways of modelling relations. Initially we think of a relation as a process which determines whether the relation exists between an input list of objects. Thus the relation 'is underneath' applied to the list [funnels, the hull] – the funnels are underneath the hull – yields an answer of 'no', but applied to [hull, funnels] – the hull is underneath the funnels – yields an answer of 'yes'. This is an example of a *decision procedure*.

In due course both functions and relations are modelled using the concepts of set theory and logic. These set models are extremely useful in mathematics and computing: indeed many mathematicians and software engineers *define* functions and relations in terms of sets.

1.2.2   A warning

In learning about sets and logic from this book you will be building mental models of the concepts. Initially, the models will be lacking in some way, rather like the model of a word processor as a typewriter and paper mentioned above. Gradually these models will be refined by further explanation and exercises. So be prepared to adapt and amend your understanding, and do not be put off by the 'brain box' who tells you that your model at any stage is wrong. Indeed, even after you have finished this book, your understanding of the basic concepts will still not be as refined as it could be; but I believe that it will prove more than adequate for most applications that you will meet.

Be warned that your understanding of the concepts *will* be limited. But do not be put off by this: most people have limited understanding of the mathematical concept of a set, for example. In particular, *infinite sets* can cause much confusion. It is for this reason that mathematicians have devised systems of *mathematical proof* which do not depend upon intuitive understanding; competent mathematicians are well aware of the limitations of their intuition. Although this book is not primarily concerned with proof, a brief introduction is given towards the end of the book and some suggestions are given for further reading. Proof *is* important in mathematics and computing, and you should certainly develop the skills. But the ability to apply mathematics is also important; the aim of this book is to give you an understanding of mathematical concepts which will be appropriate for building mathematical models.

ENJOY!

CHAPTER 2

An introduction to sets

2.1 What is a set?

The concept of 'set' is fundamental to much of modern mathematics and computer science, and so ideally we should have a clear definition. Unfortunately any attempt at a definition inevitably involves the use of the word 'set' or another word with similar meaning; in the strictest sense no definition exists! An analogous situation is the concept of 'colour': most sighted people would be unable to define the concept even though they understand it extremely well. What we need then is some intuitive understanding of the term 'set' without necessarily being able to define it.

In order to get some understanding of the notion of 'set' we can relate the concept to our existing experiences. One approach is to think of a set as a sack which contains objects (the *elements* or *members* of the set). This *model* of a set is certainly easy to use, and is one that most people can grasp. Unfortunately it can be very misleading if it is used to reason about the properties of sets or used to define other concepts. Although one should not reason intuitively, but use mathematical proof, it is still comforting to find that a proven *theorem* agrees with intuition. (That is not always the case however!) Furthermore, it will be necessary to have an understanding of sets which can help in building mathematical models of, for example, computer systems. In this book an understanding of sets is developed from experience with using computers and computer programs.

2.1.1 Set membership

In the sack model of sets, an object is said to be a member of a set if it is inside the corresponding sack. Thus if the numbers $1, 2, 3$ are placed in the sack, then 2 is a member of the set but 4 is not. Since we have decided to abandon this model, however, we need some other way of describing what *set membership* is.

Fact 2.1 A set can be defined by a procedure which calculates whether or not any given object is a member of that set. The set is said to *contain* the object.

This is somewhat oversimplifying the concept of a set, but nevertheless will serve well as a working definition. It certainly applies to *finite* sets. Note that a procedure is rather more than just a description. For example, if we try to talk of the set of 'fat people', then it is not enough simply to use the description 'fat'; something more definite is required. Thus we might use a chart, such as those commonly found in health books, to decide on the basis of weight, height, age and gender whether any person is 'fat'. Even then, we would have to specify precisely *which* chart we are going to use. Yet that will not be enough, since there will be borderline cases! A precise mathematical formulation will be required so that case an unequivocal answer will be obtained.

Now imagine that you have to specify a computer program (perhaps for a 'Speak-your-weight' machine) to decide whether a person is fat. You would need to ask just those same questions as are needed in defining the set of 'fat people'. The advantage of working with sets is that we do not become confused by issues of programming language syntax and control structures; we can focus our attention on asking those questions which are necessary to remove ambiguity and obtain precise definitions. For this reason, sets are very powerful in mathematics and computing.

There is, however, a small price to pay: we must learn a small amount of mathematical notation. We shall begin by introducing the notation to represent set membership.

Notation 2.2 If an object x is a member (an element) of a set A, then we write $x \in A$; if it is not a member then we write $x \notin A$.

Thus we can regard a set A as corresponding to a procedure which can take an object x as input, and produces an output of $x \in A$ or $x \notin A$. (It should be stressed that the truth is somewhat more complex than this when dealing with infinite sets.) A set is therefore defined by some criterion which enables us to determine whether any given object is a member of that set.

One common way of expressing this criterion is in the form of an explicit list; the procedure which corresponds to the set compares the input object with each element of the list in turn until either a match occurs (the object is a member of the set) or the end of the list is reached (the object is not a member of the set). When a set is expressed in this way we refer to the expression as a *set enumeration* or sometimes as *set display*.

Definition 2.3 A set enumeration consists of a list enclosed between braces, $\{\dots\}$.

Examples 2.4 To which of the following sets does the number 2 belong?

1. $\{1, 2, 3\}$
2. $\{4, 5, 6\}$
3. $\{3, 2, 1\}$
4. $\{one, two, three\}$

Solution 2.5 The number 2 belongs to all the sets except $\{4, 5, 6\}$. Note that the *number* 2 can be represented mathematically in many different ways, such as: 2, $1 + 1$ and two.

You may disagree with the statement that $2 \in \{\text{one}, \text{two}, \text{three}\}$. According to our definitions, the mathematical labels 2 and two refer to the same entity, and therefore 2 matches two in the set enumeration. This example serves to highlight the importance of removing ambiguity by being precise. It also illustrates the way in which more than one piece of notation may refer to the same object.

Notice how different set enumerations may refer to the same set.

Fact 2.6 Re-ordering and repeating the members in a set enumeration does not alter the set to which it refers.

Again this illustrates the way in which we may have more than one notation to represent the same idea or entity.

Examples 2.7 Which of the following set enumerations represent the same set as $\{1, 2, 3\}$?

1. $\{3, 2, 1\}$
2. $\{1, 1, 2, 3\}$
3. $\{1, 3\}$
4. $\{1, 2, 3, 4\}$
5. $\{\{1\}, \{2\}, \{3\}\}$

Solution 2.8 $\{1, 2, 3\}$ refers to the same set as $\{3, 2, 1\}$ and $\{1, 1, 2, 3\}$ but not the same as the others.

- $2 \in \{1, 2, 3\}$ but $2 \notin \{1, 3\}$
- $4 \notin \{1, 2, 3\}$ but $4 \in \{1, 2, 3, 4\}$
- Note that, for example, the *number* 2 does not match the *set* $\{2\}$ which contains the number 2, and so $2 \in \{1, 2, 3\}$ but $2 \notin \{\{1\}, \{2\}, \{3\}\}$.

Notation 2.9 Suppose that we have two set enumerations which refer to sets A and B. Then if the two enumerations always agree on whether an object x is a member of A and B, then we say that the sets A and B are *equal* and indicate this as $A = B$. For example, $\{1, 2, 3\} = \{3, 2, 1\}$. Otherwise we say that the sets are not equal and indicate this as $A \neq B$. For example $\{\{1\}, \{2\}, \{3\}\} \neq \{1, 2, 3\}$.

In the enumeration $\{\{1\}, \{2\}, \{3\}\}$, the items in the list are themselves sets; the situation is similar to the use of lists of lists, or arrays of arrays, in programming. The distinction between $\{1, 2, 3\}$ and $\{\{1\}, \{2\}, \{3\}\}$ may seem subtle, but it is very important that you understand it. If you do not understand this distinction, then not only will you have trouble understanding discrete mathematics, but you will also have trouble with data structures in computing generally.

A useful convention

In talking about sets it is often useful to be able to refer to a set containing members whose values are unspecified, $\{a, b, c\}$, for example.

Now suppose we ask whether $\{a, b, c\} = \{a, b\}$. The answer depends on whether or not c represents the same object as either a or b; if it does then the answer is 'yes', otherwise it is 'no'. For example, if $a = 1, b = 2, c = 3$ then $\{a, b, c\} \neq \{a, b\}$ but if $a = 1, b = 2, c = 2$ then $\{a, b, c\} = \{a, b\}$. If we want $\{a, b, c\}$ and $\{a, b\}$ to refer to different sets then strictly it is necessary to stipulate that a, b, c are distinct objects. In this book I shall adopt the convention that unless otherwise stated the items in a set enumeration such as $\{a, b, c\}$ are all distinct.

Definition 2.10 $\{\}$ contains no elements. It is the set enumeration for the *empty set* or *null set*.

Notation 2.11 The special symbol, $\varnothing$, is often used to indicate the empty set.

Note that $\varnothing$ does not represent the number 0. (It is sometimes said to be the Danish letter $\varnothing$ in honour of the mathematician Abel.) Unfortunately it looks similar to the special symbol for nought which it is sometimes necessary to use when writing character strings by hand (to avoid confusion with the capital letter 'O'). When writing numbers, however, the numeral 0 should *never* be written as a crossed out nought.

To confuse matters, mathematicians sometimes *define* the number 0 to be the empty set $\varnothing$; in this book I have chosen *not* to adopt that definition, and have chosen, quite deliberately, to make a distinction between 0 and $\varnothing$. It is important that you too make the same distinction; if you do not then you will soon get terribly lost, unless you are already a highly competent mathematician.

2.1.2 The nature of sets

We have introduced the notion of a set in terms of a special relation known as set membership, denoted by $\in$. If an object x is related to another object A under this relation we write $x \in A$. The object A is said to be a set; the object x is an element of that set. A set is thus a *single object*. At first it might seem strange to think of a set in this way; many mathematicians objected when the idea was first introduced at the end of the nineteenth century! Yet it has proved to be extremely useful in both mathematics and computing. For example, in a relational database, it is not allowed to have more than one object at the intersection of a row with a column; but this one object may be a set (which itself may have several elements of course).

Examples 2.12 Which of the following statements are true?

1. $0 \in \{0, 1, 2\}$
2. $\{0\} \in \{0, 1, 2\}$
3. $\{\{0\}, \{2\}, \{4\}\} = \{0, 2, 4\}$

4. $\{4, 0, 2, 0\} = \{0, 2, 4\}$
5. $\{\{4, 0, 2, 0\}\} = \{\{0, 2, 4\}\}$
6. $\varnothing \in \{\}$

Solution 2.13

1. $0 \in \{0, 1, 2\}$ is true: the element 0 matches the first element in the list used for the set enumeration.
2. $\{0\} \in \{0, 1, 2\}$ is false: the element $\{0\}$ is a set and does not match any of the numbers $0, 1, 2$.
3. $\{\{0\}, \{2\}, \{4\}\} = \{0, 2, 4\}$ is false: the elements of $\{\{0\}, \{2\}, \{4\}\}$ are all sets but those of $\{0, 2, 4\}$ are all numbers.
4. $\{4, 0, 2, 0\} = \{0, 2, 4\}$ is true: although the lists used in the set enumerations are different they will always give the same matches when checking for set membership.
5. $\{\{4, 0, 2, 0\}\} = \{\{0, 2, 4\}\}$ is true: each set contains just one element, either $\{4, 0, 2, 0\}$ or $\{0, 2, 4\}$; but we have just shown that these are equal, and so the sets containing these elements are equal.
6. $\varnothing \in \{\}$ is false: $\varnothing$ and $\{\}$ are two different ways of representing the empty set. But by definition no object can be an element of the empty set, and in particular the empty set cannot be a member of itself.

2.2   Relations between sets

2.2.1   Subset

Definition 2.14 The set A is said to be a *subset* of the set B if every element of A is also an element of B.

Notation 2.15 The subset relation is denoted by $\subseteq$; thus if A is a subset of B then we write $A \subseteq B$. For example $\{1, 2\} \subseteq \{4, 2, 3, 1\}$.

We can test whether a finite set A is a subset of B, by taking each element of A in turn (perhaps using the list in a set enumeration) and checking whether it is an element of B; this process is repeated until either an element is found which is not in B (in which case $A \nsubseteq B$) or until we have exhausted all the elements of A (in which case $A \subseteq B$).

Examples 2.16 Decide which of the following sets are subsets of $\{4, 5, 6, 7\}$:

1. $\{4, 5\}$
2. $\{5, 7, 6\}$
3. $\{4, 5, 6, 7, 8\}$
4. $\varnothing$

5. $\{4,5,6,7\}$

Solution 2.17

1. $\{4,5\} \subseteq \{4,5,6,7\}$. Using the list of the set enumeration we first check 4 then 5; both are elements of $\{4,5,6,7\}$.
2. $\{5,7,6\} \subseteq \{4,5,6,7\}$. In this case we check 5, 7 and 6.
3. $\{4,5,6,7,8\} \not\subseteq \{4,5,6,7\}$. The last item checked, 8, is not an element of $\{4,5,6,7\}$. Note though that it is the case that $\{4,5,6,7\} \subseteq \{4,5,6,7,8\}$.
4. $\varnothing \subseteq \{4,5,6,7\}$. In our subset procedure we reach the condition of having exhausted all the elements of the empty set, $\{\}$ without finding one which is not an element of $\{4,5,6,7\}$; thus $\varnothing$, that is $\{\}$, is a subset.
5. $\{4,5,6,7\} \subseteq \{4,5,6,7\}$. The term 'subset' should not be interpreted with its everyday meaning.

Fact 2.18 Every set has both the empty set ($\varnothing$ or $\{\}$) and itself as subsets.

2.2.2 Equality

Equality tends to be a notion which is taken for granted in elementary mathematics; usually there is some vague idea that two things are equal if they are the same or have the same value. If we apply such a vague notion to the two sets $\{7,8\}$ and $\{8,8,7\}$, however, we might have some difficulty in deciding whether or not they are equal. Luckily we already have a working definition of what it means to equate two sets expressed as set enumerations (see *Notation 2.9*), and so we can say that $\{7,8\} = \{8,8,7\}$.

Later we shall be meeting ways of representing sets other than using set enumerations, and indeed of representing infinite sets. Our working definition of equality based on set membership using set enumerations could be modified to say that two sets are equal if and only if they have the same members. But an alternative, and perhaps more satisfactory, approach is to define equality in terms of the subset relation.

Definition 2.19 Two sets A and B are equal if, and only if,

1. $A \subseteq B$
2. $B \subseteq A$

Applying this definition to the two sets $\{7,8\}$ and $\{8,8,7\}$ we see that they are indeed equal.

Notation 2.20 If sets A and B are equal then we write $A = B$. Thus, for example, $\{7,8\} = \{8,8,7\}$.

Quite probably you are beginning to think that this is all too obvious to warrant discussion: 'surely everyone knows that = is the symbol for equals'. In fact what most people know is the symbol for equality between numbers, which is a rather different thing to equality between sets. To be strictly correct we should use different symbols for the two types of equality.

Examples 2.21 In which of the following pairs does the first set equal the second?

1. $\{1, 2, 3\}, \{3, 2, 1\}$
2. $\{1, 2\}, \{1, 2, 3\}$
3. $\{0, 0, 1\}, \{0, 1, 1\}$
4. $\{\}, \{0\}$
5. $\{\}, \{\{\}\}$

Solution 2.22

1. $\{1, 2, 3\} = \{3, 2, 1\}$.
2. $\{1, 2\} \subseteq \{1, 2, 3\}$ but $\{1, 2, 3\} \not\subseteq \{1, 2\}$ (because 3 is not a member of $\{1, 2\}$) so $\{1, 2\} \neq \{1, 2, 3\}$.
3. $\{0, 0, 1\}, \{0, 1, 1\}$ are both equal to $\{0, 1\}$.
4. $\{\} \subseteq \{0\}$ but $\{0\} \not\subseteq \{\}$ (because the number 0 is not a member of $\{\}$) so $\{\} \neq \{0\}$.
5. $\{\} \subseteq \{\{\}\}$ but $\{\{\}\} \not\subseteq \{\}$ because the element $\{\}$ of the set $\{\{\}\}$ is not a member of the set $\{\}$ (which contains no members). Hence $\{\} \neq \{\{\}\}$.

2.2.3   Proper subset

If $A \subseteq B$ then it might also be the case that $B \subseteq A$ (in which case the two sets would be equal). Sometimes it is necessary to stipulate that A and B are not equal; that A 'really is' a subset of B.

Definition 2.23 If $A \subseteq B$ but $B \not\subseteq A$ then A is said to be a *proper subset* of B.

Notation 2.24 A proper subset is indicated by using the symbol $\subset$; thus $A \subset B$ indicates that A is a proper subset of B.

In a sense the concept of proper subset is more complex than that of subset: in order to show that $A \subset B$ it is necessary to show both $A \subseteq B$ and $B \not\subseteq A$. Perhaps for this reason, the concept is not so commonly used as that of subset.

Examples 2.25 List all the proper subsets of $\{4, 5, 6\}$.

Solution 2.26 The proper subsets of $\{4, 5, 6\}$ can be obtained systematically by removing one or more elements from the set enumeration: $\{5, 6\}$, $\{4, 6\}$, $\{4, 5\}$, $\{4\}$, $\{5\}$, $\{6\}$ and $\{\}$.

Warning! Some older books use $\subset$ to indicate a subset where we use $\subseteq$.

Examples 2.27 Which of the following statements are true?

1. $\{-1, -2\} \subseteq \{-2, -1, 0\}$
2. $\{\} \subseteq \{\}$
3. $\varnothing \subseteq \varnothing$
4. $\{\} \subset \{\}$
5. $\{\} \subseteq \{\{\}\}$
6. $\{\{\}\} \subseteq \{\}$
7. $\varnothing = \{\varnothing\}$

Solution 2.28

1. $\{-1, -2\} \subseteq \{-2, -1, 0\}$ is true. Both -1 and -2 are elements of the set $\{-2, -1, 0\}$.
2. $\{\} \subseteq \{\}$ is true. The empty set is a subset of every set, including itself (as here).
3. This is the same question as the previous one, except that the empty set has been represented by its special symbol $\varnothing$ rather than as the set enumeration $\{\}$. The answer is again true.
4. $\{\} \subset \{\}$ is false. No set can be a proper subset of itself, since by necessity it is equal to itself.
5. $\{\} \subseteq \{\{\}\}$ is true. The null set (empty set) is a subset of every set.
6. $\{\{\}\} \subseteq \{\}$ is false. The null set is an element of $\{\{\}\}$ but cannot be an element of itself, $\varnothing$.
7. $\varnothing = \{\varnothing\}$ is false since, using the alternative notation for the empty set, we have just shown that $\{\{\}\} \not\subseteq \{\}$, that is $\{\varnothing\} \not\subseteq \varnothing$.

2.3   Operations on sets

Sets can be used as input to a process which calculates another set or a number; such a process is said to be an operation on the input sets. Set operations are important not only in mathematics but also in computing; they can, for example, be used to explain or define operations in programming languages and relational databases.

   If an operation takes a single set as input it is said to be *unary*; if it takes two sets as input it is said to be *binary*. We shall first look at unary operations and then at binary operations.

2.3.1   Cardinality

Cardinality is a unary operation and gives an integer as output.

Definition 2.29 The *cardinality* of a finite set is the number of elements in the set.

It may seem odd that we should choose to invent a fancy word like 'cardinality' to mean something as simple as 'number of elements'. It is necessary to do so, however, because the concept of cardinality is really more complicated than the definition just given; in particular, it does not really make sense to talk of the number of elements in an infinite set. Our definition will, however, be adequate as a working definition.

Notation 2.30 The cardinality operator is denoted by $\#$; thus the cardinality of a set A is written as $\#A$. For example, $\#\{2, 7, 9\} = 3$.

Note that it is necessary to distinguish between the number of elements in the set and the number of elements in the list used for a set enumeration, since the list may contain repetitions. Thus $\#\{7, 2, 2, 9, 7, 2\} = 3$ even though the list itself contains six elements.

Examples 2.31 Evaluate each of the following:

1. $\#\{4, 5, 6\}$
2. $\#\{4, 6, 6, 7, 4\}$
3. $\#\{0\}$
4. $\#\{\}$

Solution 2.32

1. $\#\{4, 5, 6\} = 3$
2. $\#\{4, 6, 6, 7, 4\} = 3$
3. $\#\{0\} = 1$
4. $\#\{\} = 0$

Expressed in everyday language, both $\{0\}$ and $\{\}$ could be loosely described as 'the set containing nothing'; but the two sets are very different. This is one example in which natural language descriptions can be misleading – in this case the word 'nothing' can be interpreted either as 'the number zero' or as 'no element'. The clearest way to demonstrate the difference between the two sets is to use the cardinalities. Since $\#\{0\}$ is not equal to $\#\{\}$ then it is not possible for the two sets $\{0\}$ and $\{\}$ to be equal.

Fact 2.33 If two sets have different cardinalities then they cannot be equal.

The danger of confusing the two sets can be avoided completely by thinking *formally*, that is thinking in terms of strings of characters, rather than thinking in terms of natural language descriptions. An important aim of this book is to develop your confidence and ability in thinking formally.

2.3.2   Power set

Another unary operator is the *power set* operator. In this case the output is a set.

Definition 2.34 For any set A, the power set is the set of all subsets of A.

Notation 2.35 The power set operator is $\mathbb{P}$; the power set of A is written $\mathbb{P}A$.

In writing down the set enumeration of a power set, it is useful to have some sort of systematic procedure. One such procedure is: start by writing down the set given; then write down all the proper subsets (by leaving out elements from the original set) to form a list of all the possible subsets; finally enclose this list in braces $\{\ldots\}$ to create a single set, the power set. (An alternative approach which some books advocate is to start with the empty set $\{\}$ and build up.)

Examples 2.36 Find the following:

1. $\mathbb{P}\{4, 5, 6\}$
2. $\mathbb{P}\{1, 2\}$
3. $\mathbb{P}\{7\}$
4. $\mathbb{P}\{\}$

Solution 2.37

1. Starting with $\{4, 5, 6\}$ we can list all the possible subsets:

$$\{4, 5, 6\}, \{5, 6\}, \{4, 6\}, \{4, 5\}, \{6\}, \{5\}, \{4\}, \{\}$$

This list, however, is not the answer since we still have to form the set itself (remember that a set is regarded as a single object). The power set is obtained by placing $\{$ at the front of the list and $\}$ at the end of the list.

$$\{\{4, 5, 6\}, \{5, 6\}, \{4, 6\}, \{4, 5\}, \{6\}, \{5\}, \{4\}, \{\}\}$$

2. The subsets of $\{1, 2\}$ are $\{1, 2\}, \{2\}, \{1\}, \{\}$ and so the corresponding power set is $\mathbb{P}\{1, 2\} = \{\{1, 2\}, \{2\}, \{1\}, \{\}\}$.
3. The subsets of $\{7\}$ are $\{7\}$ and $\{\}$ so $\mathbb{P}\{7\} = \{\{7\}, \{\}\}$.
4. The only subset of $\{\}$ is $\{\}$. Enclosing this list in braces gives $\{\{\}\}$ for $\mathbb{P}\{\}$.

The cardinality and power set operators can be combined as the following examples show.

Examples 2.38 Find the following:

1. $\#\,\mathbb{P}\{4, 5, 6\}$
2. $\#\,\mathbb{P}\{1, 2\}$
3. $\#\,\mathbb{P}\{7\}$
4. $\#\,\mathbb{P}\{\}$

Can you find a pattern in these answers?

Solution 2.39

1. $\#\,\mathbb{P}\{4,5,6\} = \#\{\{4,5,6\},\{5,6\},\{4,6\},\{4,5\},\{6\},\{5\},\{4\},\{\}\} = 8$
2. $\#\,\mathbb{P}\{1,2\} = \#\{\{1,2\},\{2\},\{1\},\{\}\} = 4$
3. $\#\,\mathbb{P}\{7\} = \#\{\{7\},\{\}\} = 2$
4. $\#\,\mathbb{P}\{\} = \#\{\{\}\} = 1$

All the answers are powers of 2. In fact, if A contains n elements then $\mathbb{P}A$ contains 2^n elements.

Fact 2.40 $\#(\mathbb{P}A) = 2^{\#A}$.

This is indeed the reason for the name power set; sometimes the power set of A is written 2^A. The cardinality property of a power set can be used to check that all the elements of the power set have been listed; for example, we can be more confident that $\mathbb{P}\{\} = \{\{\}\}$ since we expect $\#\,\mathbb{P}\{\} = 2^0 = 1$.

The power set operation can be applied to any set, including another power set.

Examples 2.41 Find $\mathbb{P}\,\mathbb{P}\{7\}$.

Solution 2.42 Since $\mathbb{P}\{7\} = \{\{7\},\{\}\}$ we expect to find $2^2 = 4$ subsets of $\mathbb{P}\{7\}$. They are: $\{\{7\},\{\}\}$; $\{\{\}\}$; $\{\{7\}\}$; and $\{\}$. Thus

$$\mathbb{P}\,\mathbb{P}\{7\} = \{\{\{7\},\{\}\},\{\{\}\},\{\{7\}\},\{\}\}$$

An alternative view of the subset relation
If A is a subset of B then we may write $A \subseteq B$; but an alternative is to write $A \in \mathbb{P}B$, that is, A is an element of the power set of B. Thus to test whether A is a subset of B, we can form the power set of B and test whether A is an element of $\mathbb{P}B$. This approach does have certain advantages as the following example shows.

Examples 2.43 Is the number 5 a subset of $\{3,4,5,6\}$? Express your answer using appropriate mathematical notation.

Solution 2.44 It would seem reasonable to argue that since the number 5 is not a set it cannot be a subset of $\{3,4,5,6\}$. This conclusion is indeed correct. (Unlike some mathematicians I have chosen *not* to define the number 5 as a set; the argument I have given is therefore consistent with *my* definitions.)

We might be tempted to write $5 \nsubseteq \{3,4,5,6\}$; but strictly speaking this is meaningless since $\nsubseteq$ and $\subseteq$ denote relations between sets. There is no such problem however with writing $5 \notin \mathbb{P}\{3,4,5,6\}$, and this is how our conclusion should be expressed mathematically.

The difficulty encountered in this last example highlights a common misuse of mathematical symbols in which the symbols are treated merely as a shorthand for natural language words and phrases.

Fact 2.45 Symbols in mathematical formalism are generally best thought of in terms of objects, operations and relations.

From this point of view mathematical notation and programming languages are just different aspects of the same thing.

2.3.3   Union

Set union is an example of a binary operation which gives a set as the result.

Definition 2.46 The *union* of two sets combines all their elements into one set.

Notation 2.47 The symbol for the union operator is $\cup$. The union of sets A and B is denoted by $A \cup B$.

For example, if $A = \{1, 2, 3\}$ and $B = \{3, 4, 5\}$ then $A \cup B = \{1, 2, 3, 4, 5\}$; the elements 1 and 2 are in A, the elements 4 and 5 are in B while the element 3 is in both A and B. One way of getting the result is to combine the lists in the set enumerations to get $\{1, 2, 3, 3, 4, 5\}$. Now repetitions of the same element can be removed; in this case 3 occurs twice so the above expression can be simplified to $\{1, 2, 3, 4, 5\}$.

Fact 2.48 Writing a binary operator between the two objects on which it acts is usually called the *infix* convention. The plus sign in ordinary arithmetic is normally written using the infix convention: for example, $2 + 3$.

Definition 2.49
The definition of union can be expressed more mathematically as $x \in A \cup B$ if and only if either $x \in A$ or $x \in B$ or possibly both.

Examples 2.50 Evaluate the following:

1. $\{5, 6, 7\} \cup \{8, 9, 0\}$
2. $\{1, 2\} \cup \{1, 2, 3\}$
3. $\{4, 0\} \cup \{\{4\}, \{0\}\}$
4. $\{\} \cup \{\{\}\}$

Solution 2.51

1. $\{5, 6, 7\} \cup \{8, 9, 0\} = \{5, 6, 7, 8, 9, 0\}$.
2. $\{1, 2\} \cup \{1, 2, 3\} = \{1, 2, 3, 1, 2\} = \{1, 2, 3\}$. In fact this is a special case of the general rule that if $A \subseteq B$ then $A \cup B = B$. (In this case $\{1, 2\} \subseteq \{1, 2, 3\}$.)
3. $\{4, 0\} \cup \{\{4\}, \{0\}\} = \{4, 0, \{4\}, \{0\}\}$. There are no repeated elements.
4. $\{\} \cup \{\{\}\} = \{\{\}\}$. This is another example of the rule that if $A \subseteq B$ then $A \cup B = B$.

Fact 2.52

- If $A \subseteq B$ then $A \cup B = B$.
- Since the empty set is a subset of every set, it follows that $\varnothing \cup X = X$, no matter what the set X is.

Such general results can sometimes help simplify expressions involving set operations.

2.3.4 Intersection

Set intersection is another example of a binary operation which gives a set as the result.

Definition 2.53 The *intersection* of two sets is the set all of whose elements are in both the original sets.

Notation 2.54 The intersection operator is denoted by $\cap$; the intersection of two sets A and B is written as $A \cap B$.

For example, if $A = \{1, 2, 3, 4\}$ and $B = \{3, 4, 5, 6\}$ then $A \cap B = \{3, 4\}$. This result can be found from the following process: take each element of the first set (A) and test to see if it is also a member of the second set (B); if so then write that element down; enclose the list of the elements found in braces to form the set enumeration for $A \cap B$.

Definition 2.55 The definition of intersection can be expressed more mathematically as $x \in A \cap B$ if, and only if, both $x \in A$ and $x \in B$.

Examples 2.56 Evaluate the following:

1. $\{5, 6, 7\} \cap \{8, 9, 0\}$
2. $\{1, 2\} \cap \{1, 2, 3\}$
3. $\{4, 0\} \cap \{\{4\}, \{0\}\}$
4. $\{\} \cap \{\{\}\}$

Solution 2.57

1. $\{5, 6, 7\} \cap \{8, 9, 0\} = \{\}$ since none of $5, 6$ or 7 are elements of $\{8, 9, 0\}$.
2. $\{1, 2\} \cap \{1, 2, 3\} = \{1, 2\}$. In fact this is a special case of the general rule that if $A \subseteq B$ then $A \cap B = A$. (In this case $\{1, 2\} \subseteq \{1, 2, 3\}$.)
3. $\{4, 0\} \cap \{\{4\}, \{0\}\} = \{\}$.
4. $\{\} \cap \{\{\}\} = \{\}$. This is another example of the rule that if $A \subseteq B$ then $A \cap B = A$.

Fact 2.58

- If $A \subseteq B$ then $A \cap B = A$.
- Since the empty set is a subset of every set, it follows that $\varnothing \cap X = \varnothing$, no matter what the set X is.

2.3.5 Complements and differences

Set difference is yet another example of a binary operation which gives a set as the result.

Definition 2.59 The *set difference* of two sets A and B is the set of elements of A which are not in B. (Sometimes the term *relative complement* is used instead.)

Notation 2.60 The set difference of A and B is denoted by $A \setminus B$. (An alternative notation is to write $A - B$.)

One possible process for finding the set difference is to work through the members of the second set and check whether each is a member of the first set; when it is, then remove this element from the first set. For example, if $A = \{1, 2, 3, 4\}$ and $B = \{3, 4, 5, 6\}$ then to evaluate $A \setminus B$ we test each of the elements $3, 4, 5, 6$ in B for membership of A; 3 and 4 are members of A and so are removed from A to give $A \setminus B = \{1, 2\}$. A similar process, but using the elements of A to remove elements of B, gives $B \setminus A = \{5, 6\}$. In fact the same elements, 3 and 4, are used in the removal.

Examples 2.61 Find the following:

1. $\{2, 6, 8, 9\} \setminus \{4, 5, 6, 7, 8\}$
2. $\{4, 5, 6, 7, 8\} \setminus \{2, 6, 8, 9\}$
3. $\{7, 3, 5\} \setminus \{4, 2, 6\}$
4. $\{7, 3, 5\} \setminus \{7, 3, 5\}$

Solution 2.62

1. $\{2, 6, 8, 9\} \setminus \{4, 5, 6, 7, 8\} = \{2, 9\}$ which is obtained by removing 6 and 8 from $\{2, 6, 8, 9\}$.
2. $\{4, 5, 6, 7, 8\} \setminus \{2, 6, 8, 9\} = \{4, 5, 7\}$ which is obtained by removing 6 and 8 from $\{4, 5, 6, 7, 8\}$.
3. $\{7, 3, 5\} \setminus \{4, 2, 6\} = \{7, 3, 5\}$ since the two sets have no elements in common.
4. $\{7, 3, 5\} \setminus \{7, 3, 5\} = \varnothing$ since the two sets are identical.

Fact 2.63 In general, in evaluating $X \setminus Y$ and $Y \setminus X$ the elements removed are those of $X \cap Y$.

2.3.6 Cartesian product

This is yet another example of a binary operator. Before we define the operator however we must define a special type of object.

Definition 2.64 An *ordered pair* consists of two objects listed inside parentheses (round brackets). For example $(3, 8)$.

Ordered pairs can be thought of as coordinates like those used in plotting graphs. Thus $(3, 8)$ could be plotted with x-coordinate of 3 and y-coordinate of 8, while $(8, 3)$ could be plotted with x-coordinate of 8 and y-coordinate of 3. Note that the ordered pairs $(3, 8)$ and $(8, 3)$ are different; the order in the list is important in an ordered pair, unlike the order of elements in a set enumeration.

Ordered pairs and sets are different types of object, so great care must be taken not to confuse the concepts. But beware! Some authors regard the ordered pair (x, y) as simply an alternative notation for the set $\{x, \{x, y\}\}$; this is known as the Kuratowski definition of an ordered pair. In contrast I have chosen to make a clear distinction between ordered pairs and sets, and you too must keep that distinction. In any case, it is *always* true that (x, y) and $\{x, y\}$ are different: you should take great care to use the correct brackets for ordered pairs and for sets.

Unfortunately another source of confusion for beginners is that parentheses are also used to show the order in which operations are to be applied. For example in ordinary arithmetic, $(2 + 3) * 7$ means that the addition $+$ is carried out first while in $2 + (3 * 7)$ the multiplication $*$ is carried out first. (The use of parentheses for ordering operations will be considered in more depth in Chapter 3.)

One way of avoiding confusion is to use an alternative way of writing ordered pairs. Alternative notations that are sometimes used include:

1. The use of acute brackets $<\ldots, \ldots>$;
2. The maplet notation $\ldots \mapsto \ldots$.

For example the ordered pair $(7, 6)$ could be alternatively written as $<7, 6>$ or as $7 \mapsto 6$.

However after a little experience, the use of parentheses for ordered pairs does not cause any real difficulty.

Examples 2.65 Which of the following pairs of objects are the same?

1. $\{3, 2\}$ and $(3, 2)$
2. $\{3, 2\}$ and $\{2, 3\}$
3. $(2, 3)$ and $(3, 2)$
4. $\{(3, 2), (2, 3)\}$ and $(\{3, 2\}, \{2, 3\})$

Solution 2.66

1. $\{3, 2\}$ and $(3, 2)$ are not the same since one is a set of numbers while the other is an ordered pair of numbers.
2. $\{3, 2\}$ and $\{2, 3\}$ are the same since the order of elements in a set enumeration is not significant.
3. $(2, 3)$ and $(3, 2)$ are different since the order of the coordinates in an ordered pair is significant.
4. $\{(3, 2), (2, 3)\}$ and $(\{3, 2\}, \{2, 3\})$ are not the same: the first is a set of ordered pairs, while the second is an ordered pair of sets.

We say that braces $\{\ldots\}$ *delimit* a set while parentheses $(\,,\,)$ delimit an ordered pair.

Definition 2.67 The *Cartesian product* of two sets A and B is the set of all ordered pairs in which the first *coordinate* comes from A and the second coordinate comes from B. The term *direct product* is sometimes used. Often we speak simply of the product of A and B.

Notation 2.68 The product of A and B is denoted as $A \times B$.

Of course the symbol $\times$ is often used to indicate multiplication of *numbers*. It is important to realize that the operation of set product is radically different to number multiplication. To stress this difference, the symbol $*$ will be used to indicate the product of numbers, while the symbol $\times$ will be used to indicate the product of sets.

Nevertheless the reader may well find that in many books the same symbol is used for these two very different operations. In such cases, the intended operation can readily be determined from the nature of objects it is being applied to.

In calculating a product of sets, take each element in turn from the first set and combine it in an ordered pair with each element in turn from the second set. For example, if $A = \{1, 2\}$ and $B = \{3, 4\}$ then $A \times B = \{(1, 3), (1, 4), (2, 3), (2, 4)\}$.

Examples 2.69 Evaluate the following:

1. 2×3
2. $3 * 2$
3. $\{2\} \times \{3\}$
4. $\{3\} \times \{2\}$
5. $\{5, 6, 7\} \times \{3, 4\}$

Solution 2.70

1. $2 \times 3 = 6$. Since 2 and 3 are numbers the ordinary arithmetical product, $2 * 3$, must be meant.
2. $3 * 2 = 6$. This is 'of course' the same value as $2 * 3$.
3. $\{2\} \times \{3\} = \{(2, 3)\}$. We are now dealing with sets, so the Cartesian product is intended.
4. $\{3\} \times \{2\} = \{(3, 2)\}$. This is NOT the same as $\{2\} \times \{3\}$.
5. $\{5, 6, 7\} \times \{3, 4\} = \{(5, 3), (5, 4), (6, 3), (6, 4), (7, 3), (7, 4)\}$.

Fact 2.71 If set A and set B are different then $A \times B \neq B \times A$.

A useful property of the product set is

Fact 2.72 $\#(A \times B) = (\#A) * (\#B)$.

Indeed, it is this connection with number multiplication that gives set product its name and symbol. The property can also be used to check that the correct number of ordered pairs has found when working out an explicit example.

Note that we have used parentheses to determine the order in which the cardinality operator should be applied (rather than to delimit ordered pairs).

Examples 2.73 Calculate

1. $\#(\{2\} \times \{3\})$
2. $\#(\{3\} \times \{2\})$
3. $\#(\{5, 6, 7\} \times \{3, 4\})$

Solution 2.74

1. $\#(\{2\} \times \{3\}) = (\#\{2\}) * (\#\{3\}) = 1 * 1 = 1$. This agrees with the cardinality $\{(2, 3)\}$, which has just one element, namely $(2, 3)$.
2. $\#(\{3\} \times \{2\}) = 1$.
3. $\#(\{5, 6, 7\} \times \{3, 4\}) = 6$.

Examples 2.75 Evaluate each of the following:

1. $\#(\{\{1, 2\}, \{1\}, \{2\}\})$
2. $\mathbb{P}\{\{1\}, \{2\}\}$
3. $\{\{1\}, \{2\}\} \cup \{\{1, 2\}, \{1\}, \{2\}\}$
4. $\{\{1\}, \{2\}\} \cap \{\{1, 2\}, \{1\}, \{2\}\}$
5. $\{\{1\}, \{2\}\} \times \{\{1, 2\}, \{1\}, \{2\}\}$

Solution 2.76

1. $\#\{\{1, 2\}, \{1\}, \{2\}\} = 3$ since there are three different elements, namely $\{1, 2\}, \{1\}$ and $\{2\}$.
2. $\mathbb{P}\{\{1\}, \{2\}\} = \{\{\{1\}, \{2\}\}, \{\{2\}\}, \{\{1\}\}, \{\}\}$ which is obtained by writing down a list inside braces, $\{\ldots\}$, of four copies of $\{\{1\}, \{2\}\}$ with zero then one then two elements removed.
3. $\{\{1\}, \{2\}\} \cup \{\{1, 2\}, \{1\}, \{2\}\} = \{\{1, 2\}, \{1\}, \{2\}\}$. Note that $\{\{1\}, \{2\}\}$ is a subset of $\{\{1, 2\}, \{1\}, \{2\}\}$; the union of any set X with one of its subsets will always equal X.
4. $\{\{1\}, \{2\}\} \cap \{\{1, 2\}, \{1\}, \{2\}\} = \{\{1\}, \{2\}\}$
   since $\{\{1\}, \{2\}\} \subseteq \{1, 2\}, \{1\}, \{2\}$.
5. $\{\{1\}, \{2\}\} \times \{\{1, 2\}, \{1\}, \{2\}\}$ is a set with six members: $\{(\{1\}, \{1, 2\}),$
   $(\{1\}, \{1\}), (\{1\}, \{2\}), (\{2\}, \{1, 2\}), (\{2\}, \{1\}), (\{2\}, \{2\})\}$.

2.3.7   Generalized operations

Binary operations can usually be generalized to apply to a set of sets. Thus we can have generalized union, denoted by $\bigcup$, or generalized intersection, denoted by $\bigcap$. Their application is best explained by illustration.

Examples 2.77 Evaluate

1. $\bigcup\{\{1,2,3\},\{1,5\},\{2,3\},\{0\}\}$
2. $\bigcap\{\{1,2,3\},\{1,2,5\},\{1,2,3,5\},\{1,2,4\}\}$

Solution 2.78

1. $\bigcup\{\{1,2,3\},\{1,5\},\{2,3\},\{0\}\}$
   $= \{1,2,3\} \cup \{1,5\} \cup \{2,3\} \cup \{0\} = \{0,1,2,3,5\}$
   The generalized union is a set, each of whose elements occurs in at least one of the sets $\{1,2,3\}$, $\{1,5\}$, $\{2,3\}$ and $\{0\}$.
2. $\bigcap\{\{1,2,3\},\{1,2,5\},\{1,2,3,5\},\{1,2,4\}\}$
   $= \{1,2,3\} \cap \{1,2,5\} \cap \{1,2,3,5\} \cap \{1,2,4\} = \{1,2\}$
   The generalized intersection is a set, each of whose elements occurs in every one of the sets $\{1,2,3\}$, $\{1,2,5\}$, $\{1,2,3,5\}$ and $\{1,2,4\}$.

2.4   Case study: project teams

This book is essentially about sets, so more applications will be considered in later chapters. Nevertheless, some idea of the ways in which sets may be applied can be had by considering a simple example. This example will subsequently form the basis of a case study which will be used throughout the book to illustrate some of the ideas we shall be meeting. This case study also serves to provide examples of sets other than numbers! (In learning about sets it is useful to use numbers, since they are familiar and help to keep explanations short and simple; nevertheless, real applications of sets will necessarily include non-numerical values.)

Suppose that in a certain organization each employee is assigned to one or more project teams. Then the workforce and the project teams can be represented by sets. For example:

$$
\begin{aligned}
\textit{Workforce} &= \{Elma, Rajesh, Mary, Carlos, Mike\} \\
\textit{project\_A} &= \{Rajesh, Mike, Carlos\} \\
\textit{project\_B} &= \{Mary, Rajesh, Mike, Elma\} \\
\textit{project\_C} &= \{Elma, Carlos\} \\
\textit{project\_D} &= \{Rajesh, Mary, Mike, Carlos\}
\end{aligned}
$$

Certain information is lost by using sets; for example, there is no indication of who may be the project leaders. (To capture this missing information requires using ideas met later in this book, such as functions.) Nevertheless, there is still a fair amount of information that can be extracted using the set operations and relations outlined in this chapter.

- The number of people in each project team is given by the set cardinality operator, #: number of people in *project_A* is #{*Rajesh, Mike, Carlos*} = 3.
- The people *not* in a particular project team is given by the difference between *Workforce* and the set corresponding to the team: the set of people not on *project_D* is

$$Workforce \setminus project\_D = \{Elma\}$$

- The people working on both of two separate projects is given by set intersection: the set of people working on both *project_A* and *project_B* is

$$project\_A \cap project\_B = \{Rajesh, Mike\}$$

- The set of all possible project teams is given by $\mathbb{P}$ *Workforce*. In fact this would include the possibility of a team without any members, corresponding to the empty set. (A variation of the $\mathbb{P}$ operator is the $\mathbb{P}_1$ operator which excludes the possibility of the empty set.)
- The set of the project teams currently set up is a subset of $\mathbb{P}$ *Workforce*:

$$project\_teams = \{project\_A, project\_B, project\_C, project\_D\}$$

where, of course, each element of the set is itself a set. If the organization has a policy that everyone should be included in at least one project team, this can also be expressed using generalized union:

$$\bigcup project\_teams = Workforce$$

must always be true. An expression like this is an example of an *invariant* – something that is always true of a system.

2.5   Exercise

1. Which of the following statements are true?
    - (a) $7 \in \{0, 1, 2\}$
    - (b) $0 \in \{0, 1, 2\}$
    - (c) $\{0\} \in \{0, 1, 2\}$
    - (d) $\varnothing \in \{0, 1, 2\}$
    - (e) $\{0\} \in \{\{0, 1, 2\}\}$

(f) $\{0\} \in \{\{0\}, \{1\}, \{2\}\}$

2. Which of the following statements are true?

   (a) $\{0\} \subseteq \{0, 1, 2\}$

   (b) $\{0, 1, 2\} \subseteq \{0, 1, 2\}$

   (c) $\{2, 1\} \subseteq \{0, 1, 2\}$

   (d) $\{\} \subseteq \{0, 1, 2\}$

   (e) $\{0\} \subset \{0, 1, 2\}$

   (f) $\{0, 1, 2\} \subset \{0, 1, 2\}$

   (g) $\{2, 1\} \subset \{0, 1, 2\}$

   (h) $\varnothing \subset \{0, 1, 2\}$

   (i) $\{1, 2\} \subseteq \{\{0, 1, 2\}\}$

   (j) $\{2\} \subseteq \{\{0\}, \{1\}, \{2\}\}$

   (k) $\{0, 1, 2\} = \{0, 1, 2\}$

   (l) $\{0, 1, 2, 2\} = \{0, 1, 2\}$

   (m) $\{0, 1, 2, 0\} = \{2, 1, 0, 1\}$

   (n) $\{0, 1\} = \{0, 1, 2, 0, 1\}$

   (o) $\{\{0\}, \{1\}, \{2\}\} = \{\{0, 1, 2\}\}$

   (p) $\{\{0, 1, 2\}, \{0\}\} = \{\{0, 1, 2\}\}$

3. Evaluate each of the following:

   (a) $\#\{8, 12, 34\}$

   (b) $\#\{8, 12, 34, 12\}$

   (c) $\#\{\{8, 12, 34\}\}$

   (d) $\#\{\{8\}, \{12\}, \{34\}\}$

   (e) $\#\{\{8, 12, 34\}, \{8\}, \{12\}, \{34\}\}$

4. Evaluate each of the following:

   (a) $\mathbb{P}\{9\}$

   (b) $\mathbb{P}\,\mathbb{P}\{9\}$

   (c) $\mathbb{P}\{3, 5\}$

   (d) $\mathbb{P}\{3, 5, 3\}$

5. Evaluate each of the following:

   (a) $\{9, 5\} \cup \{5, 3\}$

   (b) $\{9, 5, 2, 6\} \cup \{1, 3, 4\}$

   (c) $\{9, 2\} \cup \{2, 9\}$

    (d) $\{\{9\}, \{2\}\} \cup \{\{9\}, \{1\}\}$

    (e) $\{9, \{2\}\} \cup \{\{9\}, 1\}$

    (f) $(\{9, 3\} \cup \{3, 4, 5\}) \cup \{1, 2, 3\}$

    (g) $\{9, 3\} \cup (\{3, 4, 5\} \cup \{1, 2, 3\})$

  Comment on these last two answers.

6. Evaluate each of the following:

    (a) $\{9, 5\} \cap \{5, 3\}$

    (b) $\{9, 5, 2, 6\} \cap \{1, 3, 4\}$

    (c) $\{9, 2\} \cap \{2, 9\}$

    (d) $\{\{9\}, \{2\}\} \cap \{\{9\}, \{1\}\}$

    (e) $\{9, \{2\}\} \cap \{\{9\}, 1\}$

    (f) $(\{9, 3\} \cap \{3, 4, 5\}) \cap \{1, 2, 3\}$

    (g) $\{9, 3\} \cap (\{3, 4, 5\} \cap \{1, 2, 3\})$

  Comment on these last two answers.

7. Evaluate each of the following:

    (a) $\{9, 5\} \setminus \{5, 3\}$

    (b) $\{9, 5, 2, 6\} \setminus \{1, 3, 4\}$

    (c) $\{9, 2\} \setminus \{2, 9\}$

    (d) $\{\{9\}, \{2\}\} \setminus \{\{9\}, \{1\}\}$

    (e) $\{9, \{2\}\} \setminus \{\{9\}, 1\}$

    (f) $(\{9, 3\} \setminus \{3, 4, 5\}) \setminus \{1, 2, 3\}$

    (g) $\{9, 3\} \setminus (\{3, 4, 5\} \setminus \{1, 2, 3\})$

  Comment on these last two answers.

8. Evaluate each of the following:

    (a) $\{9, 5\} \times \{5, 3\}$

    (b) $\{9, 2\} \times \{2, 9\}$

    (c) $\{\{9\}, \{2\}\} \times \{\{9\}, \{1\}\}$

    (d) $\{9, \{2\}\} \times \{\{9\}, 1\}$

    (e) $(\{9, 3\} \times \{3\}) \times \{1, 2\}$

    (f) $\{9, 3\} \times (\{3\} \times \{1, 2\})$

  Are the last two results the same?

9. Evaluate each of the following:

    (a) $\# \, \mathbb{P}\{3, 5\}$

    (b) $\#(\{9, 5\} \cup \{5, 3\})$

(c) $\#(\{9,5\} \cap \{5,3\})$

(d) $\#(\{9,5\} \setminus \{5,3\})$

(e) $\#(\{9,5\} \times \{5,3\})$

10. Verify by taking suitable examples that if X and Y are sets then:

    (a) $\#(X \times Y) = \#X * \#Y$

    (b) $\# \mathbb{P} X = 2^{\#X}$

    (c) $\#(X \cup Y) = \#X + \#Y - \#(X \cap Y)$

    (d) $\#(X \setminus Y) = \#X - \#(X \cap Y)$

    You may find it helpful to use small sets of no more than four elements. Note also that your results simply verify the properties listed and do not constitute mathematical proofs. In particular, care would need to be taken when deciding whether the results are applicable to infinite sets.

11. Evaluate each of the following and note which expressions give the same values.

    (a) $(\{9,3\} \cup \{3\}) \cap \{1,2,9\}$

    (b) $\{9,3\} \cup (\{3\} \cap \{1,2,9\})$

    (c) $(\{9,3\} \cap \{1,2,9\}) \cup (\{9,3\} \cap \{1,2,9\})$

    (d) $(\{9,3\} \cup \{3\}) \cap (\{9,3\} \cup \{1,2,9\})$

    (e) $\mathbb{P}(\{9,3\} \cup \{1\})$

    (f) $\mathbb{P}(\{9,3\}) \cup \mathbb{P}(\{1\})$

    (g) $\mathbb{P}(\{9,3\} \cap \{3\})$

    (h) $\mathbb{P}(\{9,3\}) \cap \mathbb{P}(\{3\})$

    (i) $\mathbb{P}(\{9,3\} \times \{1\})$

    (j) $\mathbb{P}(\{9,3\}) \times \mathbb{P}(\{1\})$

    Why has it been necessary to use brackets in these questions?

12. Suppose $E = \{-2, -1, 0\}$ and $F = \{1, 2\}$.

    (a) Evaluate:

        i. $\#(E \times F)$

        ii. $\#(F \times E)$

        iii. $\#(F \times (E \cup F))$

        iv. $\#(F \times (E \times F))$

        v. $\#(\mathbb{P} F)$

        vi. $\#(\mathbb{P} E)$

        vii. $\#(\mathbb{P}(E) \times \mathbb{P}(F))$

        viii. $\#(\mathbb{P}(E \times F))$

    ix. $\#((\mathbb{P}F) \times (\mathbb{P}(E \cup F)))$

Rewrite the expressions in this question but with as many brackets taken out as possible without causing ambiguity.

(b) Which of the following statements are true?

    i. $2 \in E$

    ii. $E \in F$

    iii. $E \subset F$

    iv. $E \subseteq F$

    v. $\varnothing \subseteq E$

    vi. $E \subseteq E$

    vii. $E \subseteq \mathbb{P}E$

    viii. $E \in \mathbb{P}E$

    ix. $\varnothing \subseteq \mathbb{P}E$

    x. $\varnothing \in \mathbb{P}E$

    xi. $\#\varnothing \in E$

13. If $C = \{\clubsuit, \diamondsuit, \heartsuit, \spadesuit\}$ and $D = \{\flat, \natural, \sharp\}$ find each of the following sets:

    (a) $C \times D$

    (b) $D \times D$

    (c) $D \times C$

    (d) $(C \times D) \times D$

    (e) $C \times (D \times D)$

14. Evaluate:

    (a) $\#\varnothing$

    (b) $\#\mathbb{P}\varnothing$

    (c) $\#\mathbb{P}\mathbb{P}\varnothing$

    (d) $\#\mathbb{P}\mathbb{P}\mathbb{P}\varnothing$

    (e) $\#\{\varnothing\}$

    (f) $\#(\{\varnothing\} \cup \{\{\varnothing\}\})$

15. Which of the following statements are true?

    (a) $\#\varnothing = 0$

    (b) $\varnothing = 0$

    (c) $\varnothing = \{0\}$

    (d) $\varnothing = \{\varnothing\}$

    (e) $\varnothing \in (\mathbb{P}\varnothing)$

(f) $\varnothing \subseteq (\mathbb{P}\varnothing)$

(g) $\#\varnothing \in \varnothing$

(h) $\varnothing \times X = \varnothing$ for any arbitrary set X

(i) $\varnothing \cap \{\varnothing\} = \varnothing$

16. Suppose that X, Y and Z are any arbitrary sets.   Which of the following statements are true?

   (a) If $X \subseteq Y$ and $Y \subseteq Z$ then $X \subseteq Z$

   (b) If $X \neq Y$ and $Y \neq Z$ then $X \neq Z$

   (c) $X \subseteq X$

   (d) If $X \subseteq Y$ and $Y \subseteq X$ then $X = Y$

   (e) $\#(X \cup Y) = \#X + \#Y$

17. The *symmetric difference* operator, $\triangle$, on two sets is the set of all the elements which are in exactly one set; elements of both sets are not included. More formally we can define $A \triangle B$ to be equal to $(A \cup B) \setminus (A \cap B)$. For example $\{1, 2, 3\} \triangle \{2, 3, 4\} = \{1, 4\}$. Evaluate

   (a) $\{0, 2, 4\} \triangle \{1, 2, 3, 4\}$

   (b) $\{0\} \triangle \{1, 2, 3, 4\}$

   (c) $\mathbb{P}\{0, 2\} \triangle \mathbb{P}\{1, 2\}$

18. Evaluate:

   (a) $\bigcup\{\{2\}, \{2, 3\}, \{2, 3, 4\}\}$

   (b) $\bigcap\{\{2\}, \{2, 3\}, \{2, 3, 4\}\}$

   (c) $\bigcup\{\{5\}, \{2, 3\}, \{3, 6\}\}$

   (d) $\bigcap\{\{5\}, \{2, 3\}, \{3, 6\}\}$

   (e) $\bigcup\mathbb{P}\{5, 3\}$

   (f) $\bigcap\mathbb{P}\{5, 3\}$

19. Suppose we have (as given in section 2.4)

$$
\begin{aligned}
\textit{Workforce} &= \{\textit{Elma}, \textit{Rajesh}, \textit{Mary}, \textit{Carlos}, \textit{Mike}\} \\
\textit{project\_A} &= \{\textit{Rajesh}, \textit{Mike}, \textit{Carlos}\} \\
\textit{project\_B} &= \{\textit{Mary}, \textit{Rajesh}, \textit{Mike}, \textit{Elma}\} \\
\textit{project\_C} &= \{\textit{Elma}, \textit{Carlos}\} \\
\textit{project\_D} &= \{\textit{Rajesh}, \textit{Mary}, \textit{Mike}, \textit{Carlos}\}
\end{aligned}
$$

Evaluate each of the following:

   (a) $\#\textit{Workforce}$

   (b) $\textit{Workforce} \setminus \textit{project\_A}$

(c) *project_C* $\cap$ *project_D*

(d) *project_A* $\cup$ *project_B*

(e) *project_A* $\times$ *project_C*

(f) *project_C* $\times$ *project_A*

(g) $\mathbb{P}$ *project_C*

(h) $\mathbb{P}_1$ *project_C*

(i) $\bigcup${*project_A, project_B, project_C*}

(j) $\bigcap${*project_A, project_B, project_C*}

CHAPTER 3

Propositional logic

3.1 What is logic?

A scan through the literature will show that the term 'logic' is used in several different ways reflecting a wide variety of uses. It is often said that logic began in ancient Greece, and in particular with the work of Aristotle. This 'classical logic' is based on ordinary language and is concerned with certain rules by means of which correct conclusions may be made from a given set of facts. It is worth exploring briefly some of the other ways in which logic is used, and how these are related. Some of the ideas will be useful for developing a model of logic as used in this book, where it is used as a means of deciding set membership. The reader will almost certainly have come across, or will come across, some of these various uses.

3.1.1 Symbolic logic

This is a development of classical logic in which symbols are used instead of or in addition to ordinary language in presenting an argument. A basic idea is that of a *proposition*. A proposition can be defined as a statement which can be either *true* or *false*. For example

- 'the cat sat on the mat'
- 'all dogs like bones'

are both examples of propositions.

Now it is possible to form *compound propositions* from simpler ones by the use of words and phrases like 'and', 'or', 'not', 'only if', 'if and only if'. For example

- 'either the cat sat on the mat or all dogs like bones'.

However such expressions are often ambiguous. Does 'or' in the previous example exclude the possibility that 'the cat sat on the mat' and 'all dogs like bones' are both true, or is this possibility included? In ordinary language both the *inclusive* and *exclusive* meanings of 'or' may be found. To overcome this problem, special symbols can be introduced known as *connectives*. Thus the symbol $\vee$ is used to represent 'or' in the inclusive sense.

31

In discussing arguments it is convenient to use letters such as p, q, r to stand for propositions. Thus the example given above is of the form $p \lor q$ where p stands for 'the cat sat on the mat' and q stands for 'all dogs like bones'. Note that we have assumed that the inclusive meaning of 'or' is intended. An expression like $p \lor q$ is called a *propositional form*; it represents the way in which the proposition has been built from simpler ones.

Similarly each of the following compound propositions can be written using symbols:

- The statement 'not all dogs like bones' can be written as $\neg$'all dogs like bones'.
  This has the form $\neg p$.
- The statement 'all dogs like bones and the cat sat on the mat' can be written as 'all dogs like bones'$\land$'the cat sat on the mat'.
  This has the form $p \land q$.
- The statement 'the cat sat on the mat only if all dogs like bones' can be written as 'the cat sat on the mat'$\Rightarrow$'all dogs like bones'.
  This has the form $p \Rightarrow q$.
- The statement 'the cat sat on the mat if and only if all dogs like bones' can be written as 'the cat sat on the mat'$\Leftrightarrow$'all dogs like bones'.
  This has the form $p \Leftrightarrow q$.

Later we shall be using these connectives in a more formal way. Most beginners in logic prefer to have some intuitive notion of what the connectives 'really mean' in terms of everyday language, as given in the list above. Although this can be reassuring, it can also be *highly* misleading and confusing, especially for $\Rightarrow$. If possible, it is better to think about connectives as symbols which have certain properties and obey certain rules. This is all part of the more abstract, more formal approach which you will need to develop in thinking about computers and software. Sadly there is no easy route to this kind of thinking!

3.1.2   Formal logic

This is very important in software engineering. Instead of propositions written in ordinary language, strings of symbols can be written down using a prescribed set of symbols. Furthermore in any given logical system, only certain strings are acceptable, known as *well formed formulae* or *wff*s; they correspond to the statements or propositions of classical and symbolic logic. (Formal logic can be regarded as a development of symbolic logic, and indeed the two terms are often used synonymously.) A programming language is one example of a *formal logic*; another example is ordinary arithmetic. For example the string of characters between the quotes in '$2+3 = 5$' is a *wff*. Another example of a *wff* is '$2+3 = 78$'. An example of a string of characters which is not a *wff* is '$= + = 2 =$'.

In any logical system, certain *wff*s are taken to be *theorems*; the theorems of a system can be regarded as those statements which are true. Examples of theorems in arithmetic include '$2 + 3 = 5$' and '$7 - 6 = 1$' but not '$2 + 3 = 78$', even though all three are *wff*s.

3.1.3   Circuit logic

Digital electronics such as that used in computer hardware is based on switching various points in a circuit between a 'high' voltage and a 'low' voltage (typically 15 V and 0 V respectively). A common convention is to let 1 represent a point at a high voltage and to let 0 represent a point at a low voltage. In a computer memory, these 1's and 0's are used to store information. In processing these values, *gates* are used; these are special circuits that produce output values (1 or 0) dependent upon the input voltages. For example the OR gate has two inputs and one output whose value is 0 when both inputs are 0, but 1 otherwise. A gate thus acts as a function which 'calculates' an output value from the input values. The function can be specified by listing all the possible combinations of input and giving the output in each case. Such a listing is often referred to as a *truth table*. Typically letters such as a and b are used to represent inputs, while algebraic looking expressions represent output; for example if a and b are the inputs to an OR gate, then the output is represented as $a + b$. The OR gate has a truth table:

Truth table for the OR gate		
a	b	$a + b$
1	1	1
1	0	1
0	1	1
0	0	0

The term 'truth table' reflects the connection with more traditional uses of logic. Each input or output corresponds to a proposition. A value of 1 corresponds to a 'true' proposition and a value of 0 to a 'false' proposition. Connectives may be used to represent gates. For example $\vee$ is associated with the OR gate, so that $a + b$ in circuit logic could be written as $p \vee q$. This would, however, be highly unconventional in circuit logic!

   WARNING Do not use the notation for circuit logic when doing propositional logic.

3.2   Propositional forms

We have seen in 3.1.1 that two propositions p and q can be combined to form a more complex proposition, called a *compound proposition*. For example, using $\vee$ can give propositions like:

1. $(2 + 3 = 5) \vee (1 + 1 = 2)$

2. $(2 \in \mathbb{N}) \vee (2^3 = 7)$
3. $(2 \in \mathbb{N}) \vee ((2 + 3 = 5) \vee (1 + 1 = 2))$

Examples 3.1 For each of these three examples write down the proposition that corresponds to p and the proposition that corresponds to q in $p \vee q$.

Solution 3.2

1. p is $(2 + 3 = 5)$ and q is $(1 + 1 = 2)$.
2. p is $(2 \in \mathbb{N})$ and q is $(2^3 = 7)$.
3. p is $(2 \in \mathbb{N})$ and q is $((2 + 3 = 5) \vee (1 + 1 = 2))$.

In the last example q itself is a compound proposition.

All three propositions can be obtained by replacing p and q in $p \vee q$ by appropriate propositions; they all have the same *propositional form*. Letters such as p and q are called *propositional letters*.

In general, compound propositions can be formed from simpler ones by using special symbols known as *logical connectives*; these are $\neg, \vee, \wedge, \Rightarrow, \Leftrightarrow$. The properties of these connectives can be described by considering the corresponding propositional forms.

Examples 3.3 Substitute p by $(2 + 3 = 5)$, q by $(7 - 2 = 10)$ and r by $(1 \in \mathbb{Z})$ in each of the following propositional forms.

1. $\neg p$
2. $p \vee q$
3. $p \wedge q$
4. $p \Rightarrow q$
5. $p \Leftrightarrow q$
6. $\neg p \wedge q$
7. $\neg(p \wedge q)$
8. $\neg p \vee \neg q$
9. $(p \wedge q) \vee r$

Solution 3.4

1. $\neg(2 + 3 = 5)$
2. $(2 + 3 = 5) \vee (7 - 2 = 10)$
3. $(2 + 3 = 5) \wedge (7 - 2 = 10)$
4. $(2 + 3 = 5) \Rightarrow (7 - 2 = 10)$
5. $(2 + 3 = 5) \Leftrightarrow (7 - 2 = 10)$
6. $\neg(2 + 3 = 5) \wedge (7 - 2 = 10)$
7. $\neg((2 + 3 = 5) \wedge (7 - 2 = 10))$
8. $\neg(2 + 3 = 5) \vee \neg(7 - 2 = 10)$
9. $((2 + 3 = 5) \wedge (7 - 2 = 10)) \vee (1 \in \mathbb{Z})$

These solutions were obtained by using a common software tool: the `replace` operation on a word processor. The process is 'mechanical' and can be achieved without understanding what the symbols represent. Note that brackets were used to improve readability; the use of brackets is discussed in section 3.6.

Examples 3.5   For each of the following compound propositions write down a propositional form together with appropriate substitutions for the propositional letters which would have given the compound proposition.

1. $\neg(2 + 3 = 7)$
2. $(2 + 3 = 7) \vee (7 - 2 = 5)$
3. $\neg((2 + 3 = 7) \wedge (7 - 2 = 5))$
4. $((2 + 3 = 7) \wedge (7 - 2 = 5)) \vee (0 \in \mathbb{Z})$
5. $2 + 3 = 1 + 4 = 5$

Solution 3.6

1. $\neg p$ with p as $(2 + 3 = 7)$.
2. $p \vee q$ with p as $(2 + 3 = 7)$ and q as $(7 - 2 = 5)$.
3. $\neg p$ with p as $((2 + 3 = 7) \wedge (7 - 2 = 5))$. Note that this is not the only possible solution; $\neg(p \wedge q)$ with p as $(2 + 3 = 7)$ and q as $(7 - 2 = 5)$ is another possibility.
4. One possible solution is $(p \wedge q) \vee r$ with p as $(2 + 3 = 7)$, q as $(7 - 2 = 5)$ and r as $(0 \in \mathbb{Z})$.
5. At first glance $2 + 3 = 1 + 4 = 5$ does not seem to be a compound proposition at all. In fact it is really a shorthand way of writing

$$(2 + 3 = 1 + 4) \wedge (1 + 4 = 5)$$

Thus it can be obtained from $p \wedge q$ with p as $(2 + 3 = 5)$ and q as $(1 + 4 = 5)$. Such shorthand forms as $2 + 3 = 1 + 4 = 5$ are very convenient in many ways, but they are perhaps best avoided until you are fairly experienced in handling logic, and will not be used in general throughout this book.

Note that propositional forms are often called just 'propositions'; usually the context makes it clear whether or not it is really a propositional form.

So far we have not attached any meaning, or *semantics*, to the connective symbols. One approach to giving meaning to the symbols has already been presented briefly in section 3.1.1; connectives may be regarded as substitutes for natural language word and phrases. This 'intuitive' approach, however, has its limitations and should only ever be used as an approximate guide. Instead, we shall think about the symbols in terms of operations on truth values; this approach is similar to that used in circuit logic.

3.3 Definitions of connectives

With each propositional letter such as p, q or r we can associate a variable truth value of T ('true') or F ('false'). Each connective can then be associated with a *truth function*. A truth function takes one or more truth values as input and returns a single truth value as output; its behaviour can be defined in a *truth table*. From a purely mathematical point of view, a truth table tells us all we need to know about the connective to which it refers, and so we consider the connective to be defined by its truth table. If you feel that this is a rather limited view of logic, you are right; but to consider the matter more fully would take us too deeply into philosophy and more complex issues of logic which are perhaps best avoided in an introductory book.

3.3.1 Negation

The *negation* connective $\neg$ (pronounced NOT) is defined by the following truth table:

Negation	
p	$\neg p$
T	F
F	T

The table gives the truth values of $\neg p$ corresponding to the various truth values of p.

3.3.2 Disjunction

The *disjunction* connective $\vee$ (pronounced OR) is defined by the following truth table:

Disjunction		
p	q	$p \vee q$
T	T	T
T	F	T
F	T	T
F	F	F

The table gives the truth values of $p \vee q$ corresponding to the various possible combinations of truth values of p and q.

3.3.3 Conjunction

The *conjunction* connective $\wedge$ (pronounced AND) is defined by the following truth table:

Conjunction		
p	q	$p \wedge q$
T	T	T
T	F	F
F	T	F
F	F	F

3.3.4   Material conditional

The *conditional* connective $\Rightarrow$ (pronounced ONLY IF) can be defined by the following truth table:

Conditional		
p	q	$p \Rightarrow q$
T	T	T
T	F	F
F	T	T
F	F	T

Thus if $p \Rightarrow q$ is known to be true, then p can be true *only if* q is true.

In 3.5.3 an alternative definition is given for the $\Rightarrow$ connective.

3.3.5   Biconditional

The *biconditional* connective $\Leftrightarrow$ (pronounced IF AND ONLY IF) can be defined by the following truth table:

Biconditional		
p	q	$p \Leftrightarrow q$
T	T	T
T	F	F
F	T	F
F	F	T

In 3.5.3 an alternative definition is given for the $\Leftrightarrow$ connective.

3.4   Other propositional forms

Truth tables are also used to obtain the function which corresponds to any compound propositional form. In order to do this, it is necessary to analyse the expression for the propositional form in order to decide the order in which the connectives have been used; that is to say we need to *parse* the expression. This is best illustrated by means of some simple worked examples.

Examples 3.7   In each of the following basic forms, replace p by $(\neg p)$ and q by $(p \wedge q)$ to obtain a more complex form. Hence obtain a truth table for each of the new forms.

1. $\neg p$
2. $p \vee q$
3. $p \wedge q$
4. $p \Rightarrow q$
5. $p \Leftrightarrow q$

Solution 3.8 First we obtain the new propositional forms by substitution, for example by using the `replace` operation in a word processor.

1. $\neg(\neg p)$
2. $(\neg p) \vee (p \wedge q)$
3. $(\neg p) \wedge (p \wedge q)$
4. $(\neg p) \Rightarrow (p \wedge q)$
5. $(\neg p) \Leftrightarrow (p \wedge q)$

Now in order to find the truth table in each case we begin with the substituted terms namely $(\neg p)$ and $(p \wedge q)$. These are basic forms for which we already have the truth tables

p	q	$(\neg p)$	$(p \wedge q)$
T	T	F	T
T	F	F	F
F	T	T	F
F	F	T	F

We can now build up the more complex expressions using the truth tables of the basic forms.

1. $\neg(\neg p)$

p	$(\neg p)$	$\neg(\neg p)$
T	F	T
F	T	F

The third column has been obtained from the second column by applying the truth function corresponding to $\neg$.

2. $(\neg p) \vee (p \wedge q)$

p	q	$(\neg p)$	$(p \wedge q)$	$(\neg p) \vee (p \wedge q)$
T	T	F	T	T
T	F	F	F	F
F	T	T	F	T
F	F	T	F	T

In this case the last column has been obtained by applying the $\lor$ truth function to each pair of values from the third and fourth columns.

3. $(\neg p) \land (p \land q)$

p	q	$(\neg p)$	$(p \land q)$	$(\neg p) \land (p \land q)$
T	T	F	T	F
T	F	F	F	F
F	T	T	F	F
F	F	T	F	F

The last column has been obtained by using the $\land$ truth function to the two previous columns.

4. $(\neg p) \Rightarrow (p \land q)$

p	q	$(\neg p)$	$(p \land q)$	$(\neg p) \Rightarrow (p \land q)$
T	T	F	T	T
T	F	F	F	T
F	T	T	F	F
F	F	T	F	F

5. $(\neg p) \Leftrightarrow (p \land q)$

p	q	$(\neg p)$	$(p \land q)$	$(\neg p) \Leftrightarrow (p \land q)$
T	T	F	T	F
T	F	F	F	T
F	T	T	F	F
F	F	T	F	F

In the last set of examples it was known in advance what the basic forms were. Usually, it is necessary to work backwards from a given form.

Examples 3.9   Obtain a truth table for each of the following propositional forms:

1. $(\neg p) \land q$
2. $\neg(p \land q)$
3. $(\neg p) \lor (\neg q)$
4. $(p \land q) \lor r$
5. $(p \land (\neg q)) \Rightarrow p$

Solution 3.10

1. This expression can be obtained by replacing p by $(\neg p)$ in $p \land q$.

p	q	$(\neg p)$	$(\neg p) \land q$
T	T	F	F
T	F	F	F
F	T	T	T
F	F	T	F

2. This expression can be obtained by replacing p by $(p \wedge q)$ in $\neg p$.

p	q	$(p \wedge q)$	$\neg(p \wedge q)$
T	T	T	F
T	F	F	T
F	T	F	T
F	F	F	T

3. Substitute $(\neg p)$ for p and $(\neg q)$ for q in $p \vee q$.

p	q	$(\neg p)$	$(\neg q)$	$(\neg p) \vee (\neg q)$
T	T	F	F	F
T	F	F	T	T
F	T	T	F	T
F	F	T	T	T

4. In this case we have a third 'variable' r and so there are going to be 8 rows in the truth table.

p	q	r	$(p \wedge q)$	$(p \wedge q) \vee r$
T	T	T	T	T
T	T	F	T	T
T	F	T	F	T
T	F	F	F	F
F	T	T	F	T
F	T	F	F	F
F	F	T	F	T
F	F	F	F	F

5. In this case there are two levels of replacement. The compound proposition can be built up substituting $(\neg q)$ for q in $(p \wedge q)$ to give $(p \wedge (\neg q))$, then substituting $(p \wedge (\neg q))$ for p in $p \Rightarrow q$.

p	q	$(\neg q)$	$(p \wedge (\neg q))$	$(p \wedge (\neg q)) \Rightarrow p$
T	T	F	F	T
T	F	T	T	T
F	T	F	F	T
F	F	T	F	T

3.5   Metalanguage

3.5.1   Tautologies

The propositional form $(p \wedge \neg q) \Rightarrow p$ was shown above to be always true no matter what the truth values of p and q. This is an important property of this particular propositional form, since no matter what propositions we substitute for p and q the resultant proposition will always have a truth value of T. A form such as $(p \wedge \neg q) \Rightarrow p$ whose truth value is always T is called a *tautology*. We can denote that a particular propositional form is a tautology by using the *semantic turnstile*, $\models$.

$$\models (p \wedge \neg q) \Rightarrow p \tag{3.1}$$

Note that the expression 3.1 is *not* a proposition; it is saying something *about* a propositional form. This is sometimes a bit confusing to the beginner since it seems reasonable to describe an expression like 3.1 as a statement which can be true or false. In a naïve view of logic, a proposition is indeed described as a statement which can be true or false; however this description can be misleading. In a more formal approach to logic, a logical statement (that is a proposition) can only be built up from certain characters; by definition these characters do not include $\models$. A further point to note is that there is no truth function associated with $\models$, so $\models$ does not have a truth table and cannot be a connective. Thus an expression like (3.1) does not have associated with it a truth value of T or F, and so cannot be a proposition. (Note that to be precise we should distinguish between saying an expression has a truth value of T and saying that the expression is true!)

An expression like 3.1 is said to be a *metastatement*. In general we use *metalanguage* to talk about propositions. Words such as 'tautology' are part of this metalanguage. Associated with this metalanguage are *metasymbols* such as $\models$.

3.5.2   Equivalent propositions

Consider the truth tables of $\neg(p \vee q)$ and $(\neg p) \wedge (\neg q)$:

p	q	$(p \vee q)$	$\neg(p \vee q)$	$(\neg p)$	$(\neg q)$	$(\neg p) \wedge (\neg q)$
T	T	T	F	F	F	F
T	F	T	F	F	T	F
F	T	T	F	T	F	F
F	F	F	T	T	T	T

It can be seen that the truth values in the fourth and seventh columns are the same; that is, $\neg(p \vee q)$ and $(\neg p) \wedge (\neg q)$ have the same truth tables. We express this in our metalanguage by saying that the two propositional forms are *equivalent* and use the metasymbol $\equiv$ to indicate this.

$$\neg(p \vee q) \equiv (\neg p) \wedge (\neg q)$$

Examples 3.11   Look over the truth tables in the worked examples above and find:

1. two forms which are equivalent to p
2. one form which is equivalent to $p \Rightarrow q$

and write down these metastatements using an appropriate metasymbol.

Solution 3.12

    1. $\neg(\neg p) \equiv p$
       $(\neg p) \Rightarrow (p \wedge q) \equiv p$
    2. $(\neg p) \vee (p \wedge q) \equiv p \Rightarrow q$

3.5.3   Alternative definitions for $\Rightarrow$ and $\Leftrightarrow$

We have defined $\Rightarrow$ and $\Leftrightarrow$ by giving their truth tables; an alternative approach is to define them in terms of equivalent expressions. For example, since we have shown that

$$(\neg p) \vee (p \wedge q) \equiv p \Rightarrow q$$

we *could* define $p \Rightarrow q$ as being simply a shorthand way of writing $(\neg p) \vee (p \wedge q)$. In fact there is an even simpler, and quite common, form which is equivalent to $p \Rightarrow q$. From the truth table

p	q	$\neg p$	$\neg p \vee q$
T	T	F	T
T	F	F	F
F	T	T	T
F	F	T	T

it can be seen that

$$\neg p \vee q \equiv p \Rightarrow q$$

The expression $p \Rightarrow q$ may be regarded as an abbreviation of $\neg p \vee q$; in other words $p \Rightarrow q$ is *defined* to be equivalent to $\neg p \vee q$.

    Defining two expressions to be equivalent is rather different to defining them separately and then *discovering* they are equivalent. In mathematics, a special symbol is used to show that two expressions are being defined as equivalent; it is $\hat{=}$. Thus we write the following:

$$p \Rightarrow q \hat{=} \neg p \vee q$$

    Similarly it is possible to treat $p \Leftrightarrow q$ as an abbreviation of $(p \Rightarrow q) \wedge (q \Rightarrow p)$, which has the truth table:

p	q	$p \Rightarrow q$	$q \Rightarrow p$	$(p \Rightarrow q) \wedge (q \Rightarrow p)$
T	T	T	T	T
T	F	F	T	F
F	T	T	F	F
F	F	T	T	T

Thus we can write down the abbreviation definition,

$$p \Leftrightarrow q \hat{=} (p \Rightarrow q) \wedge (q \Rightarrow p)$$

3.5.4   Logical implication

Another important metasymbol is the *semantic turnstile* $\models$. This symbol has already been introduced as a way of expressing that a propositional form is a tautology. In fact this is just one specialized use of the symbol. More generally it is used to denote *logical implication*. (Sometimes the phrase *semantic entailment* is used.)

The following definition may seem a little complex, but hopefully the examples should make clearer what is meant.

Notation 3.13 We write

$$P_1, P_2, P_3, \ldots \models Q \tag{3.2}$$

to indicate the fact that whenever the propositional forms P_1, P_2, P_3, ... all have value T, then the propositional form Q also has value T.

For example, we have:

$$p, p \Rightarrow q \models q \tag{3.3}$$

Whenever both p and $p \Rightarrow q$ have truth value of T then q also has a value of T, as can be seen by inspection of the truth tables.

We see that p and $p \Rightarrow q$ logically imply q. The propositional forms to the left of $\models$ are called the *antecedents* and the single propositional form to the right of $\models$ is called the *consequent*.

The list of antecedents may comprise just one propositional form. One particular case of this is the truth value T. In this case the consequent must always have value T irrespective of the values of the logical variables; the consequent must be a *tautology*. For example

$$T \models (p \wedge \neg q) \Rightarrow p \tag{3.4}$$

expresses the fact that $(p \wedge \neg q) \Rightarrow p$ is a tautology. By convention we can leave out the T, and this is indeed the normal way of expressing a tautology in symbols:

$$\models (p \wedge \neg q) \Rightarrow p \tag{3.5}$$

3.5.5   Contradictions

If a propositional form always has the value F irrespective of the values of the logical variables, it is called a *contradiction*. This can be denoted by writing the contradiction as the (single) antecedent and F as the consequent. For example

$$p \wedge \neg p \models F \tag{3.6}$$

indicates that whenever $p \wedge \neg p$ has a value of T then $T = F$, that is 'true' = 'false'; this can never be the case, and so $p \wedge \neg p$ can only ever have the value F.

3.6   The use of brackets

In a strictly formal approach to logic, all the basic propositional forms include brackets:

- $(\neg p)$
- $(p \vee q)$
- $(p \wedge q)$
- $(p \Rightarrow q)$
- $(p \Leftrightarrow q)$

This ensures that propositional forms can always be interpreted unambiguously. For example suppose we wrote down

$$\neg p \wedge q$$

This could be interpreted in two ways according to whether the $\neg$ applies to just p or to the bigger expression $p \wedge q$; the truth tables are different.

- If $\neg$ applies to just p, then we would have $((\neg p) \wedge q)$ with truth table

p	q	$(\neg p)$	$((\neg p) \wedge q)$
T	T	F	F
T	F	F	F
F	T	T	T
F	F	T	F

- If $\neg$ applies to $p \wedge q$, then we would have $(\neg(p \wedge q))$ with truth table

p	q	$(p \wedge q)$	$(\neg(p \wedge q))$
T	T	T	F
T	F	F	T
F	T	F	T
F	F	F	T

Comparing the two propositional forms we can see that they are not equivalent, that is

$$((\neg p) \wedge q) \not\equiv (\neg(p \wedge q))$$

It therefore seems that we do need to put in brackets to make clear which propositional form is meant.

Examples 3.14   In the example just given, there *is* a relationship between the two alternative forms; state this relationship and express it using an appropriate metasymbol.

Solution 3.15   The truth table for $((\neg p) \wedge q)$ has only one T, in the third row; the truth table for $(\neg(p \wedge q))$ has the value of T in three rows including the third. Thus every combination of truth values of p and q which gives $((\neg p) \wedge q)$ a value of T also gives $(\neg(p \wedge q))$ a value of T. We say that

$((\neg p) \wedge q)$ logically implies $(\neg(p \wedge q))$

and write

$((\neg p) \wedge q) \models (\neg(p \wedge q))$

In general, though, there will not be any relationship between alternative forms.

Strictly speaking we should always use brackets. Unfortunately, this tends to make logic look extremely cumbersome; anyone who has seen a program written in LISP (a programming language based on logic) will have been struck by the abundance of brackets. Clearly we need some well defined rules to enable us to leave brackets out whenever we can, without causing ambiguity; luckily there are a set of rules similar to those used in ordinary arithmetic.

It should be emphasized, however, that there is nothing wrong with putting in brackets where they are not needed, and that it is better to err on the side of using too many brackets than too few.

3.6.1   Rules for dropping brackets

Outer brackets
The outermost brackets of an expression can always be dropped; but take care to put them back in when substituting a propositional letter by this form.

For example it is perfectly acceptable to write $(\neg p) \wedge q$ instead of $((\neg p) \wedge q)$ and $\neg(p \wedge q)$ instead of $(\neg(p \wedge q))$ since in each case the outer brackets are understood to be 'really' there. Now suppose we want to replace p by $\neg(p \wedge q)$ in $p \vee q$. We must put the brackets back in: $\neg(p \wedge q) \vee q$. This should be familiar to most readers from ordinary arithmetic and algebra. For example we would normally write $2 + 3$ rather than the more pedantic $(2 + 3)$. But if we want to substitute $2 + 3$ for y in $5 \times y$ we must remember that the brackets are 'really' there and write $5 \times (2 + 3)$; because the brackets are no longer the outermost ones, our rule does not permit us to leave them out.

Order of precedence
In ordinary arithmetic there is a convention that, in the absence of brackets, certain operations will always be carried out before others: powers will be evaluated first; then multiplications and divisions; then addition and subtractions. We can describe this by saying that powers have a higher priority than multiplication and division, which in turn have a higher priority than addition and subtraction. In other words the expression $1 + 2 \times 5^3$ will be *parsed* as $(1 + (2 \times (5^3)))$ and evaluates to 251.

Similar conventions have been suggested for propositional logic; unfortunately there is no universal agreement on which one! In this book, the simplest convention has been adopted, namely that $\neg$ applies to the smallest possible section of a propositional form; this will be either a single propositional letter such as q or an expression enclosed in a matching pair of brackets such as $((p \wedge q) \Rightarrow r)$. This convention leads to better readability of propositional forms; furthermore, all expressions written according to this convention will be correctly interpreted by all other conventions.

Repetitions of the same connective

An arithmetic expression like $1 + 2 + 3$ is evaluated from left to right, that is it is interpreted as an being an abbreviation for $((1 + 2) + 3)$; this is certainly how the expression would be interpreted in a program. It so happens that for repeated addition it does not matter which '+' is carried out first: $(1 + 2) + 3$ and $1 + (2 + 3)$ each have the same value of 6. Similarly for multiplication: $(2 \times 3) \times 4$ and $2 \times (3 \times 4)$ each have the same value of 24.

Operations like $+$ and $\times$ are said to be *associative*. Not all operations are associative. For example $(7 - 4) - 3$ has a value of 0 but $7 - (4 - 3)$ has a value of 6. Thus care must be taken in writing $7 - 4 - 3$ to ensure that we really do mean $(7 - 4) - 3$. As a general rule it is safest to put in the brackets; it makes for better readability anyway.

A similar principle has been suggested for propositional logic. Thus $p \wedge q \wedge r$ would be interpreted as $(p \wedge q) \wedge r$ and similarly $p \Leftrightarrow q \Leftrightarrow r$ would be interpreted as $(p \Leftrightarrow q) \Leftrightarrow r$. But note that the usual convention for $\Rightarrow$ is the opposite: $p \Rightarrow q \Rightarrow r$ is normally interpreted as $p \Rightarrow (q \Rightarrow r)$. Since the use of such conventions may lead to confusion, it is safest always to use brackets to make the meaning clear.

Nevertheless, three of the four binary connectives are associative (the exception being the $\Rightarrow$ connective):

$$(p \vee q) \vee r \;\equiv\; p \vee (q \vee r)$$
$$(p \wedge q) \wedge r \;\equiv\; p \wedge (q \wedge r)$$
$$(p \Leftrightarrow q) \Leftrightarrow r \;\equiv\; p \Leftrightarrow (q \Leftrightarrow r)$$
$$\text{but}$$
$$(p \Rightarrow q) \Rightarrow r \;\not\equiv\; p \Rightarrow (q \Rightarrow r)$$

For associative connectives it is perfectly reasonable to leave out the brackets since there is essentially no difference between the alternative interpretations. Thus it is perfectly reasonable to write $p \vee q \vee r$, $p \wedge q \wedge r$ and $p \Leftrightarrow q \Leftrightarrow r$ without brackets, but $p \Rightarrow (q \Rightarrow r)$ and $(p \Rightarrow q) \Rightarrow r$ must be written *with* brackets.

3.7   Modelling software with logic

Modelling software with logic really requires a knowledge of predicate logic and the use of quantifiers as described in Chapter 4. Nevertheless, the reader may find it helpful to see a simple example, even if some of the details may not yet be fully understood.

3.7.1   The IF ... THEN ... control structure

Programming languages usually have a control structure like

```
IF cond THEN action
```

where cond is some *guard condition* and action is some process which takes place if the guard condition, cond, is true. It is possible to build a simple logic model for this control structure. If p corresponds to cond and q is always true after action has occurred, then the control structure can be modelled by $p \Rightarrow q$. This will be made clearer by considering some simple examples. (Note that, with these definitions, we ought to refer to p and q as *predicates*, not propositions. Predicates will be defined in the next chapter; for the present, however, we need not be concerned with this finer point.)

Example 1
Consider a program which inputs the values of two variables, x and z, recalculates z under certain conditions, then outputs the values of x and z. Suppose that z is recalculated with the line

```
IF  x > 0 THEN  z := x
```

Clearly the guard condition, cond, is x > 0 while the action is z := x.
    An intuitive approach to understanding the effect of this command line is to consider some specific input values of x and z:

- Input x as 1 and z as 2.
  In this case the guard condition is true, so the value of z is altered to x. The output values are $x = 1$ and $z = 1$. Note that the output value of z is the same as the input value of x.
- Input x as 0 and z as 3.
  In this case the guard condition is false, so the output values are $x = 0$ and $z = 3$. Note that the output value of z is not the same as the input value of x.
- Input x as -1 and z as -1.
  In this case the guard condition is again false, so the value of z remains unaltered. The output values are $x = -1$ and $z = -1$. Note that in this case, although the guard condition was false, the output values of z and x are indeed the same; this is simply because the input values were the same!

Rather than rely upon such specific examples to describe the behaviour of our program, we can instead try to write down a logical expression which *always* has a truth value of T whenever the program executes correctly. Such an expression is called an *invariant* of the program. Ideally the invariant should have a truth value of F whenever the program does not execute as we require. The invariant then enables us to check whether the program is doing what it is meant to be doing.

In order to write down the invariant, we introduce the symbol $x?$ to represent the input value of x and $z!$ to represent the output value of z. Notice the distinction between program variables, such as x and z, and mathematical variables such as $x?$ and $z!$: x and z refer to memory locations which store numerical values, while $x?$ and $z!$ refer to values held in these memory locations at specific times in the execution of the program.

Now first consider the expression $x? > 0$; for convenience refer to this expression as p. This is clearly not an invariant of the program, as can be seen by considering the examples given above ($0 > 0$ and $-1 > 0$ are both equal to F). Next consider the expression $z! = x?$; for convenience refer to this expression as q. Again, this cannot be an invariant of the program, as can be seen by considering the second example given above (for which the expression becomes $0 = 3$ which has truth value F). However, we can see from the examples, that there are three allowable combinations of truth values for p and q:

- $p = T$ and $q = T$
- $p = F$ and $q = F$
- $p = F$ and $q = T$

The combination $p = T$ and $q = F$ is not allowable; if the situation were ever to arise in which the input value of x was greater than 0 but the output value of z differed from the input value of x, then the program would not have behaved as intended. Thus the program invariant must be of the form $p \ddagger q$ where $\ddagger$ is a connective with the following truth table:

p	q	$p \ddagger q$
T	T	T
T	F	F
F	T	T
F	F	T

In fact we have already defined such a connective: the conditional connective $\Rightarrow$. Thus our required invariant is $p \Rightarrow q$, that is:

$$(x? > 0) \Rightarrow (z! = x?) \tag{3.7}$$

Expression 3.7 has the value T for all combinations of input and output which are possible for the program.

- Whenever the input value of x is greater than 0, then the output value of z will be set equal to x. Both p and q will have value T, and so by referring to the truth table we see that $p \Rightarrow r$ will also have a value of T.

- Whenever the input value of **x** is not greater than 0, then the input value of **z** will stay as the output value; the output value could *possibly* equal the input value of **x**, but not necessarily. Thus p will definitely have a value of F but the value of q could be either T or F.

However, it can be seen from the truth table that $p \Rightarrow r$ is always T whenever p is F, irrespective of the truth value of r.

Example 2
Suppose now that the program is modified such that a third variable **y** is input, and **z** is recalculated by

```
IF x > 0 THEN
     ( IF y > 0 THEN z := x )
```

There are two structures of the type IF ... THEN ... , the second being nested inside the first. This can be modelled by

$$(x? > 0) \Rightarrow ((y? > 0) \Rightarrow (z! = x?)) \tag{3.8}$$

where $y?$ represents the input value of **y**. Expression 3.8 has the form $p \Rightarrow (r \Rightarrow q)$. We can show this expression to be equivalent to the form $(p \wedge r) \Rightarrow q$ (for example, by using truth tables). This suggests a neater way of coding the program as

```
IF x > 0 AND y > 0 THEN z := x
```

(where the Boolean operator AND corresponds to the $\wedge$ connective).

3.8 Exercise

1. Construct truth tables for each of the following propositional forms. By inspection of the truth tables, write down those forms which are tautologies, those pairs of forms which are equivalent and those which are contradictory. Denote these facts using appropriate metasymbols and the truth constants T and F.

   (a) One variable:

       i. $\neg p \vee p$;
       ii. $\neg p \wedge p$;
       iii. $\neg \neg p \vee p$;
       iv. $\neg \neg p \wedge p$.

   (b) Two variables:

       i. $\neg p \vee q$;

      ii. $p \wedge \neg q$;

     iii. $\neg(p \wedge \neg q)$;

     iv. $(p \Rightarrow q) \wedge (q \Rightarrow p)$;

      v. $(p \Rightarrow q) \Leftrightarrow (q \Rightarrow p)$;

     vi. $(\neg p \vee p) \Rightarrow (\neg q \wedge q)$;

    vii. $(\neg p \wedge p) \Rightarrow (\neg q \wedge q)$.

(c) Three variables:

       i. $(p \vee q) \vee r$;

      ii. $p \vee (q \vee r)$;

     iii. $(p \wedge q) \wedge r$;

     iv. $p \wedge (q \wedge r)$;

      v. $(p \Rightarrow q) \Rightarrow r$;

     vi. $p \Rightarrow (q \Rightarrow r)$;

    vii. $(p \Leftrightarrow q) \Leftrightarrow r$;

   viii. $p \Leftrightarrow (q \Leftrightarrow r)$;

     ix. $p \Rightarrow (q \wedge r)$;

      x. $(p \Rightarrow q) \wedge (p \Rightarrow r)$;

     xi. $(p \wedge q) \Rightarrow (p \wedge r)$;

    xii. $p \Rightarrow (q \vee r)$;

   xiii. $(p \Rightarrow q) \vee (p \Rightarrow r)$;

   xiv. $p \Leftrightarrow (q \wedge r)$;

    xv. $(p \Leftrightarrow q) \wedge (p \Leftrightarrow r)$;

   xvi. $(p \wedge q) \Leftrightarrow (p \wedge r)$;

  xvii. $p \Leftrightarrow (q \vee r)$;

 xviii. $(p \Leftrightarrow q) \vee (p \Leftrightarrow r)$;

   xix. $\neg((p \wedge \neg q) \Rightarrow \neg r)$;

    xx. $(p \Rightarrow q) \Leftrightarrow (\neg q \Rightarrow \neg p)$;

   xxi. $((p \wedge q) \Rightarrow r) \Leftrightarrow ((p \Rightarrow r) \vee (q \Rightarrow r))$.

2. Demonstrate each of the following:

  (a) $\neg p \Rightarrow p \models p$;

  (b) $p \models q \Rightarrow (p \wedge q)$;

  (c) $p, p \Rightarrow q \models q$;

  (d) $p \Rightarrow q, r \Rightarrow s \models (p \vee r) \Rightarrow (q \vee s)$.

CHAPTER 4

Predicate logic

4.1   Set comprehension

So far in this book we have concentrated on small sets which we have been able to write as set enumerations. In general, however, sets may be too large to be defined in this way. What is required is some means of expressing set membership other than explicitly listing all the elements. An alternative to set enumeration is *set comprehension* in which we use a *predicate* instead of a list.

4.1.1   Predicates and free variables

The expression $2 < 5$ is a proposition; we can give it a truth value, in this case T. Similarly the expression $7 < 5$ is another proposition with truth value F. The expression $x < 5$ however cannot be given a truth value since we do not know the value of x. We say that x is a *free variable*.

Because of the free variable x, the expression $x < 5$ is not associated with a single truth value; instead it is associated with a *family of propositions*, each corresponding to a single value of x, and each having a truth value – for example:

- replacing x by 1 gives $1 < 5$ which has truth value T;
- replacing x by 2 gives $2 < 5$ which has truth value T;
- replacing x by 7 gives $7 < 5$ which has truth value F.

The expression $x < 5$ is an example of a *predicate*.

Fact 4.1 A predicate does not have a truth value.

Fact 4.2 A predicate contains one or more free variables.

In the expression $x < y$ there are two free variables, x and y; each free variable can be replaced by a constant value. For example, replacing y by 5 gives the expression $x < 5$, which is still a predicate but now with just one free variable, x. Replacing both x and y by (possibly different) constants gives a proposition; for example, replacing x by 2 and y by 3 gives the proposition $2 < 3$.

Fact 4.3 Replacing all the free variables in a predicate by constants gives a proposition (with a truth value).

Note that it is sometimes necessary to use a letter to represent a constant in which case that letter would not represent a free variable. For example, if we have defined a to have a constant value of 2, say, then the expression $a < 5$ is indeed a proposition since we have a *binding* of a to 2. Traditionally in mathematics, letters from the early part of the alphabet $(a, b, c, \ldots)$ are used to represent constants while those from the end of the alphabet $(x, y, z, \ldots)$ are used to represent variables.

Examples 4.4

1. Which of the following expressions are predicates (and what are the free variables), and which are propositions (and what are the truth values)?

    (a) $4 + 2 = 7$

    (b) $x + 2 = 7$

    (c) $x + y = z$

    (d) $2 = 2$

    (e) $x = x$

2. For each of the following predicates obtain a proposition by replacing the free variables by constant values: x by 3, y by 2 and z by 0. State the truth value of each proposition.

    (a) $x > 3$

    (b) $y = 5$

    (c) $x + y = z$

    (d) $x = x$

Solution 4.5

1. (a) $4 + 2 = 7$ is a proposition with truth value F.

    (b) $x + 2 = 7$ is a predicate with free variable x.

    (c) $x + y = z$ is a predicate with free variables x, y and z.

    (d) $2 = 2$ is a proposition with truth value T.

    (e) It may be tempting to say that since $x = x$ is obviously true for any value of x then it is a proposition; but it is in fact a predicate because it has a free variable, x. A truth value of T or F can only ever be associated with a proposition, not with a predicate, so $x = x$ does not have a truth value.

2. (a) $3 > 3, F$

    (b) $2 = 5, F$

    (c) $3 + 2 = 0, F$

    (d) $3 = 3, T$. Note that replacing x in $x = x$ by *any* constant always gives a proposition with truth value T; for example, $2 = 2, 7 = 7$ or $-1 = -1$ all of which, though different, have truth value of T.

4.1.2   Logical connectives

Logical connectives can be used to make more complex predicates from simpler ones. For example if we have the predicate

$$x^2 > 5 \wedge x < 10$$

then substituting the free variable x by a constant throughout the predicate will generate a compound proposition. So if x is replaced by 1 we obtain the proposition

$$1^2 > 5 \wedge 1 < 10$$

Now $1^2 > 5$ has truth value F while $1 < 10$ has truth value T. From the properties of $\wedge$ we can conclude that the compound proposition has truth value F.

Examples 4.6 For each of the following compound predicates produce two propositions by first substituting 2 for x then 3 for x. Determine the truth value of each proposition.

1. $x^2 > 5 \wedge x < 10$
2. $x^2 > 5 \vee x < 10$
3. $x^2 > 5 \Rightarrow x < 10$
4. $\neg\, (x^2 > 5) \wedge x < 10$

Solution 4.7 The two basic predicates $x^2 > 5$ and $x < 10$ become:

- $2^2 > 5$ and $2 < 10$ for $x = 2$ with truth values F and T;
- $3^2 > 5$ and $3 < 10$ for $x = 3$ with truth values T and T.

Thus the compound propositions (and their truth values) are:

1. $2^2 > 5 \wedge 2 < 10$ with truth value $F \wedge T = F$
   $3^2 > 5 \wedge 3 < 10$ with truth value $T \wedge T = T$;
2. $2^2 > 5 \vee 2 < 10$ with truth value $F \vee T = T$
   $3^2 > 5 \vee 3 < 10$ with truth value $T \vee T = T$;
3. $2^2 > 5 \Rightarrow 2 < 10$ with truth value $F \Rightarrow T = T$
   $3^2 > 5 \Rightarrow 3 < 10$ with truth value $T \Rightarrow T = T$;
4. $\neg\, (2^2 > 5) \wedge 2 < 10$ with truth value $\neg\, (F) \wedge T = T \wedge T = T$
   $\neg\, (3^2 > 5) \wedge 3 < 10$ with truth value $\neg\, (T) \wedge T = F \wedge T = F$.

4.1.3   Using predicates to define sets

Consider the predicate $x < 5$; it expresses a condition on the free variable x. Now this condition can be used as the criterion for deciding set membership – the corresponding set is the set of all values x for which $x < 5$. We can denote this set as $\{\, x \mid x < 5\,\}$. This notation is an example of *set comprehension*; more precisely it is an example of *untyped* set comprehension.

The vertical line $\mid$ is read as 'such that' and is placed before the constraint on the possible values of x. The notation $\{\, x \mid x < 5\,\}$ is thus read as 'the set of values x such that x is less than five'.

An interpretation of set comprehension

A set comprehension can be modelled by a decision procedure which takes input values and for each input gives a result of T or F; input values which give T are elements of the set and those which give F are not.

For example, suppose we have $\{\,x \mid x < 5\,\}$. To test whether any particular value is a member of this set we substitute that value for x in the predicate. Thus to test whether $2 \in \{\,x \mid x < 5\,\}$ we replace x by 2 in $x < 5$ to get the proposition $2 < 5$. Using our intuitive understanding of the '$<$' relation we can see that this particular proposition is true. We conclude that $2 \in \{\,x \mid x < 5\,\}$ is also true.

Of course, $2 \in \{\,x \mid x < 5\,\}$ is itself a proposition; naturally it has the same truth value as $2 < 5$.

Fact 4.8 For any proposition we can always find another proposition which takes the form of a statement about set membership and which has the same truth value.

Examples 4.9 Which of the following are elements of $\{\,x \mid x < 5\,\}$?

1. 1
2. 7
3. −1

Solution 4.10 Substituting each of $1, 7, -1$ for x in the predicate $x < 5$ gives the following propositions and truth values:

1. $1 < 5$ which has truth value T;
2. $7 < 5$ which has truth value F;
3. $-1 < 5$ which has truth value T.

Thus we see that 1 and −1 are elements of the set but 7 is not.

4.1.4   Typed predicates

In our model of set comprehension, we use a predicate to produce a truth value of T or F for each input value. Unfortunately there are a number of difficulties with the *untyped* set comprehension we have been using. These may be demonstrated by some simple problems.

Examples 4.11 Which of the following are true?

1. $\varnothing \in \{\,x \mid x < 5\,\}$
2. $2.3 \in \{\,x \mid x < 5\,\}$
3. $S \in S$ where $S = \{\,x \mid x \notin x\,\}$

Solution 4.12

1. From elementary set theory we know that $\varnothing \in \{\,x \mid x < 5\,\}$ is not true. However, substituting $\varnothing$ into $x < 5$ gives $\varnothing < 5$ which is clearly meaningless, and so we cannot determine the truth value of $\varnothing < 5$ – in that case we would not be able to determine whether or not $\varnothing$ is an element of $\{\,x \mid x < 5\,\}$.

2. Replacing x by 2.3 in $x < 5$ gives $2.3 < 5$. Is this true? The answer depends upon whether '$<$' is intended to apply to all numbers or just integers. If it applies to all numbers (or more precisely to what are called *real numbers*) then $2.3 < 5$ is true, and so 2.3 is indeed a member of the set $\{\, x \mid x < 5 \,\}$. But if $<$ applies only to integers, then we cannot determine a truth value for $2.3 < 5$.

3. If $S \notin S$ then by definition S is an element of $\{x \mid x \notin x\}$, so $S \in S$; if $S \in S$ then S cannot be an element of $\{x \mid x \notin x\}$, so $S \notin S$. Either way we end up with a contradiction. We cannot find a truth value for $S \notin S$.

One possible way round the difficulty of meaningless statements such as $\varnothing < 5$ would be to use *three value logic* in which the possible truth values are 'true', 'false' and 'undecided'; inputs giving 'true' are members of the set, while other inputs ('false' and 'undecided') are not members. However other difficulties remain.

- We would still not know whether or not $x < 5$ is intended to refer just to integers. In fact even this difficulty could be overcome by insisting that $<$ always refers to all real numbers, and by changing the predicate to read $x \in \mathbb{Z} \wedge x < 5$ if only integer values are required. Note that with three value logic we need to redefine connectives such as $\wedge$ to include combinations with 'undecided' in. Clearly however the use of three value logic is somewhat more complicated than two value logic.

- For $S = \{x \mid x \notin x\}$ we still have a contradiction: $S \notin S$ is 'false' because $S \notin S$ is 'undecided'! The difficulty really stems from the nature of the procedure associated with $x \mid x \notin x$; calling this procedure with x set to S necessitates finding the truth value of $S \notin S$; this can be determined by calling the procedure with x set to S which necessitates finding the truth value of $S \notin S$; this can be determined by calling the procedure with x set to S which necessitates finding the truth value of $S \notin S$; ... and so on *ad infinitum*. In fact we have an example of *infinite recursion* such that the procedure can never terminate. The procedure is not really a decision procedure. Since a set always corresponds to a decision procedure, then $S = \{x \mid x \notin x\}$ cannot be a set!

What is needed is a notation (and appropriate rules) to ensure that set comprehensions do indeed represent sets, and which is based upon two value logic. The basic idea we shall use is that in set comprehension we start with an object which is known to be a set, and use the predicate to form a subset of this.

Notation 4.13 Some common pre-defined sets include:

- $\varnothing$, the empty set;
- $\mathbb{N}$, the set of all natural numbers starting with $0, 1, 2, 3, \ldots$;
- $\mathbb{N}_1$, the set of all natural numbers except 0;
- $\mathbb{Z}$, the set of all integers, including zero, positive and negative values such as -1;
- $\mathbb{Q}$, the set of all rational numbers, including all integer and fractional values such as $\frac{37}{23}$;

- $\mathbb{R}$, the set of all real numbers, including rational and irrational values such as $\pi, \sqrt{2}$ (an irrational number is one that cannot be written as a fraction with two integers).

In this book we shall not be using $\mathbb{Q}$ and $\mathbb{R}$, so do not worry if you are not familiar with these concepts. We shall however be using $\mathbb{N}$ and $\mathbb{Z}$ so make sure you get to know these two sets.

Thus for example the difficulties we have in substituting values for x in $x < 5$ may be avoided if we declare that the x may only take values from a specified *given* set. This set is often referred to as the *type* of x.

Notation 4.14 If x is declared to be an element of the set X then we write $x :\in X$. For example $x :\in \mathbb{N}$ declares that x is a positive integer or zero.

Note also that some authors write $x \in \mathbb{N}$ or $x : \mathbb{N}$ where I have used $x :\in \mathbb{N}$.

Definition 4.15 A *typed* set comprehension is of the form

$$\{\, x :\in X \mid p(x) \,\}$$

where $p(x)$ is a predicate with free variable x.

For example $\{\, x :\in \mathbb{N} \mid x < 5 \,\}$ represents the set of natural numbers less than 5; since this is a finite set, we can find a corresponding set enumeration, namely $\{0, 1, 2, 3, 4\}$.

The effect of the predicate following $\mid$ is to select the required values from the pre-defined set to form a subset; in order to simplify later explanations, I shall refer to such a predicate as a *selector predicate*.

For typed set comprehension, the procedure for testing whether an object b say is a member of a set $\{\, x :\in A \mid p(x) \,\}$ is as follows:

1. Test if b is an element of the given set A; if $x \notin A$ then $b \notin \{\, x :\in A \mid p(x) \,\}$.
2. If $b \in A$ then we can replace x by b in the predicate to form a proposition; if the proposition is true then $b \in \{\, x :\in A \mid p(x) \,\}$, but if the proposition is false then $b \notin \{\, x :\in A \mid p(x) \,\}$.

Examples 4.16 Which of the following are true?

1. $2.3 \in \{\, x :\in \mathbb{N} \mid x < 5 \,\}$
2. $\varnothing \in \{\, x :\in \mathbb{N} \mid x < 5 \,\}$

Solution 4.17

1. Since $2.3 \notin \mathbb{N}$ then $2.3 \notin \{\, x :\in \mathbb{N} \mid x < 5 \,\}$;
2. Since $\varnothing \notin \mathbb{N}$ then $\varnothing \notin \{\, x :\in \mathbb{N} \mid x < 5 \,\}$.

Using these ideas it is now possible to obtain set enumerations for set comprehensions in simple cases.

Examples 4.18 Find the set enumerations for

1. $\{x :\in \mathbb{N} \mid x < 3\}$
2. $\{y :\in \mathbb{N} \mid y < 3\}$
3. $\{x :\in \mathbb{N} \mid x > 3 \wedge x < 6\}$

Solution 4.19

1. $x :\in \mathbb{N}$ chooses values of x from the set of natural numbers, that is it chooses $0, 1, 2, 3, 4, \ldots$ (though not necessarily in that order). From these, the predicate $x < 3$ selects the values $0, 1, 2$. The resulting set is therefore $\{0, 1, 2\}$.
2. $y :\in \mathbb{N}$ chooses values of y from the set of natural numbers, that is it chooses $0, 1, 2, 3, 4, \ldots$ (though not necessarily in that order). From these, the predicate $y < 3$ selects the values $0, 1, 2$. The resulting set is therefore $\{0, 1, 2\}$. Clearly this set is identical to $\{x :\in \mathbb{N} \mid x < 3\}$ – replacing x by y throughout gives another expression which represents the same object.
3. The selector predicate is a *compound* predicate made up from two simpler predicates and the logical AND connective. This compound predicate is satisfied by values of x of 4 and 5. The set enumeration is simply $\{4, 5\}$.

4.1.5   Bound variables

The expression $\{x :\in \mathbb{Z} \mid x < 5\}$ represents a particular object, namely the set of integers less than 5. Replacing x by any other variable will give an equivalent expression; for example $\{y :\in \mathbb{Z} \mid y < 5\}$ also represents the set of integers less than 5. In contrast, replacing x by a constant will result in a meaningless expression, for example $\{3 :\in \mathbb{Z} \mid 3 < 5\}$. This is in marked contrast to the predicate $x < 5$ where replacing x by another variable gives a predicate, $y < 5$ for example, while replacing x by a constant still gives a meaningful expression – a proposition in fact – for example $3 < 5$. We say that the x in $\{x :\in \mathbb{Z} \mid x < 5\}$ is a *bound* variable, while in the predicate on its own, $x < 5$, x is a free variable; the set comprehension $\{x \mid \ldots\}$ is said to *bind* the variable x.

Note that $2 \in \{x :\in \mathbb{Z} \mid x < 5\}$ is a proposition – it is effectively asserting that 2 is a number less than 5 – with truth value T.

Fact 4.20 A proposition has no free variables but may have one or more bound variables.

Examples 4.21 In each of the following expressions, is x bound or free?

1. $x^2 > 5$
2. $\{x :\in \mathbb{Z} \mid x^2 > 5\}$
3. $x \in \{y :\in \mathbb{Z} \mid y^2 > 5\}$
4. $x \in \{x :\in \mathbb{Z} \mid x^2 > 5\}$

Which expressions are predicates?

Solution 4.22

1. x is free in $x^2 > 5$, which is a predicate.
2. x is bound in $\{\, x :\in \mathbb{Z} \mid x^2 > 5 \,\}$ which represents a set.
3. Although y is bound in $x \in \{\, y :\in \mathbb{Z} \mid y^2 > 5 \,\}$ ($\{\, y :\in \mathbb{Z} \mid y^2 > 5 \,\}$ represents a particular set), the expression as a whole is a predicate with one free variable, namely x.
4. In the subexpression, $\{\, x :\in \mathbb{Z} \mid x^2 > 5 \,\}$, x is bound by the set comprehension. The subexpression represents a constant value, in this case a set. This set could equally as well be represented as $\{\, y :\in \mathbb{Z} \mid y^2 > 5 \,\}$, and the whole expression as $x \in \{\, y :\in \mathbb{Z} \mid y^2 > 5 \,\}$; we have just concluded that in this latter expression x is free. We therefore have the interesting situation that in the expression

$$x \in \{\, x :\in \mathbb{Z} \mid x^2 > 5 \,\}$$

the first occurrence of x is free, but the subsequent occurrences are bound. We can replace the free occurrence of x by a constant, to obtain a proposition, but not the bound occurrences. Thus $2 \in \{\, x :\in \mathbb{Z} \mid x^2 > 5 \,\}$ is a proposition associated with the predicate $x \in \{\, x :\in \mathbb{Z} \mid x^2 > 5 \,\}$, whereas the expression $2 \in \{\, 2 :\in \mathbb{Z} \mid 2^2 > 5 \,\}$ is meaningless.

Bound variables and loops
The set enumeration for $\{x :\in A \mid p(x)\}$ can be evaluated whenever the pre-defined set A is *finite*. The variable x is allowed to range through all the possible values in A; whenever $p(x)$ evaluates to T, that value of x is added to the set enumeration. The bound variable corresponds to the loop variable that would be used in an actual program.

4.2   Set replacement

An extension to set comprehension is to follow the declaration and predicate by a formula. For example, $\{\, x :\in \mathbb{N} \mid x < 5 \bullet x^2 \,\}$ refers to the set of numbers which are squares of numbers less than five, that is $\{0, 1, 4, 9, 16\}$. Each element in the set $\{\, x :\in \mathbb{N} \mid x < 5 \,\}$ is *replaced* by its square, so that a new, *replacement* set $\{\, x :\in \mathbb{N} \mid x < 5 \bullet x^2 \,\}$ is formed.

Examples 4.23 If $A = \{x :\in \mathbb{Z} \mid x < 7 \bullet x - 2\}$ which of the following are elements of A?

1. 4
2. 1
3. -3

4. 6
5. 3.6

Solution 4.24

1. We need to find an integer b less than 7 such that $4 = b - 2$ is true; the only possible value for B is 6. Hence $4 \in \{x :\in \mathbb{Z} | x < 7 \bullet x - 2\}$.
2. $1 = 3 - 2$ and $3 \in \{x :\in \mathbb{Z} \mid x < 7\}$ so $1 \in \{x :\in \mathbb{Z} \mid x < 7 \bullet x - 2\}$.
3. $-3 = -1 - 2$ so $-3 \in A$.
4. The only value of b for which $6 = b - 2$ is 8; but $8 \not< 7$ so it follows that $8 \notin \{x :\in \mathbb{Z} \mid x < 7\}$. Hence 6 is not an element of A.
5. The only value of b for which $3.6 = b - 2$ is 5.6, but this value is not an integer. Hence although $5.6 < 7$, because $5.6 \notin \mathbb{Z}$ we conclude that $5.6 \notin \{x :\in \mathbb{Z} | x < 7\}$, and hence that $3.6 \notin \{x :\in \mathbb{Z} | x < 7 \bullet x - 2\}$

4.2.1 Finding set enumerations for finite sets

In an expression of the form

$$\{ x :\in A \mid p(x) \bullet \mathcal{T}(x) \}$$

values of x which belong to the set A are 'piped' to the selector predicate $p(x)$; each of the values selected by $p(x)$ is then 'piped' to a further process $\mathcal{T}(x)$ through the $\bullet$ symbol. A useful mnemonic for the $\bullet$ symbol is to think of it as a fat pipe seen end on – the selected values are passed down the pipe to the following procedure; the thin pipe $\mid$ is used to pass values to a selector predicate only. This approach can be useful to obtain the set enumeration of a finite set.

Examples 4.25 Find set enumerations for each of the following:

1. $\{ x :\in \mathbb{N} \mid x < 5 \bullet 2 * x \}$
2. $\{ x :\in \mathbb{N} \mid x < 4 \bullet x^2 \}$
3. $\{ x :\in \mathbb{N} \mid x < 3 \bullet (x, x^2) \}$
4. $\{ x, y :\in \mathbb{N} \mid x < 3 \wedge y < 2 \bullet (x, y) \}$
5. $\{ x, y :\in \mathbb{N} \mid x < 3 \wedge y < x \bullet (x, y) \}$

Solution 4.26

1. As before, $x :\in \mathbb{N} \mid x < 5$ produces the numbers $0, 1, 2, 3, 4$. Each of these numbers is doubled to give the replacement set $\{0, 2, 4, 6, 8\}$.
2. The selector predicate is $x < 4$ which chooses the values $0, 1, 2, 3$ only; squaring these gives the set $\{0, 1, 4, 9\}$.
3. $x :\in \mathbb{N} \mid x < 3$ selects the values $0, 1, 2$; each of these is further processed by the replacement term in $\bullet$ (x, x^2) to give an ordered pair: for example 2 gives $(2, 4)$. The resulting set is therefore a set of ordered pairs, namely $\{(0, 0), (1, 1), (2, 4)\}$.

4. In this case we have *two* variables declared, x and y. The predicate $x < 3 \wedge y < 2$ selects combinations of x and y; in this case x can take values of $0, 1, 2$ and y can take values of $0, 1$. These values are then processed by $\bullet$ (x, y) to give ordered pairs. The resulting set is

$$\{(0, 0), (1, 0), (2, 0), (0, 1), (1, 1), (2, 1)\}$$

5. The $y < x$ part of the selector predicate means that the values of y cannot be chosen independently of x; for example when $x = 1$ we can only choose values of y less than 1 (in fact there is only one possible value for y when $x = 1$, namely $y = 0$). The set is $\{(1, 0), (2, 0), (2, 1)\}$.

Note that predicates such as $x < 3 \wedge y < 2$ and $y < x$ are examples of *two place* predicates.

Examples 4.27 Express the following sets using set comprehension, if necessary using the $\bullet$ extension.

1. $\{0, 3, 6\}$
2. $\{\{0\}, \{1\}\}$

Solution 4.28 *Possible* answers are

1. $\{0, 3, 6\} = \{\, x :\in \mathbb{N} \mid x < 3 \bullet 3 * x \,\}$
2. $\{\{0\}, \{1\}\} = \{\, x :\in \mathbb{N} \mid x < 2 \bullet \{x\} \,\}$

4.3   Venn diagrams

4.3.1   Visual representation of logic

During the eighteenth and nineteenth centuries, various attempts were made to use some form of diagram to help with understanding logic; names associated with these attempts include Euler, Karnaugh and Lewis Carroll. The most widely used kind of diagram has been that devised by Venn, and named after him. It should be stressed that the diagrams themselves are not important; although many people find them useful as an aid, you do not *have* to use them.

Suppose we want a visual model of the propositional form $p \wedge q$. This can be achieved by considering the compound *predicate* $p(x) \wedge q(x)$ where:

- x is chosen from $\mathcal{X}$, the set of points which lie inside a box; $x :\in \mathcal{X}$. See Figure 4.1.
- $p(x)$ is the predicate 'x is inside $\mathcal{P}$', where $\mathcal{P}$ is a region inside the box; $\mathcal{P} \subseteq \mathcal{X}$. See Figure 4.2.
- $q(x)$ is the predicate 'x is inside $\mathcal{Q}$', where $\mathcal{Q}$ is a region inside the box; $\mathcal{Q} \subseteq \mathcal{X}$. See Figure 4.3.

Figure 4.1 *The region* $\mathcal{X}$.

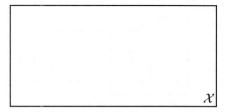

Figure 4.2 *The region* $\mathcal{P}$ *for which* $p(x) = T$.

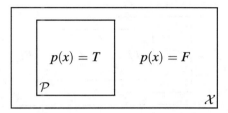

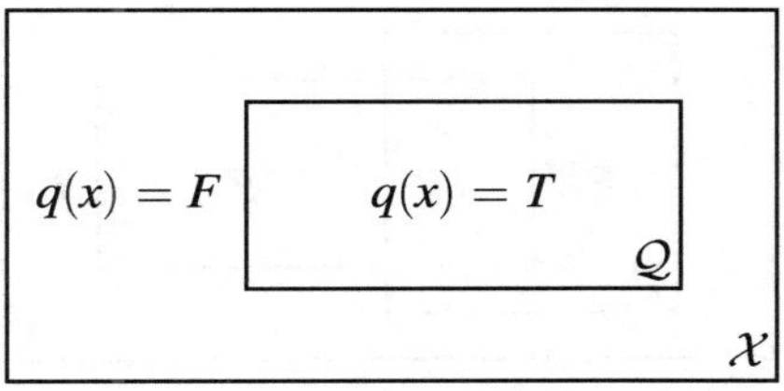

Figure 4.3 *The region* $\mathcal{Q}$ *for which* $q(x) = T$.

Combining the diagrams for $p(x)$ and $q(x)$ we get Figure 4.4.

Now consider the points e, f, g and h. For a point such as e which lies inside both circles, $p(e)$ and $q(e)$ both have truth value equal to T; similar results can be obtained for the other points. Thus we can build up a table:

$p(e) = T$ $q(e) = T$
$p(f) = T$ $q(f) = F$
$p(g) = F$ $q(g) = T$
$p(h) = F$ $q(h) = F$

Thus with each of the four distinct regions inside the box, we can associate a different pair of truth values for p and q, as shown in Figure 4.5; furthermore, by using the properties of $\wedge$ we can associate a truth value for $p \wedge q$ with each region, as shown in Figure 4.6.

The Venn diagram can thus be used as an alternative to truth tables for dis-

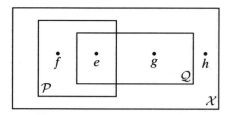

Figure 4.4 *Partition of X into four areas.*

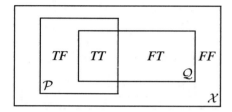

Figure 4.5 *Combinations of truth values in the four areas of X.*

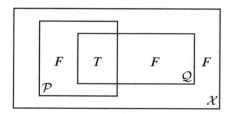

Figure 4.6 *Truth values of p ∧ q.*

playing the truth values of a propositional form; each region of the Venn diagram corresponds to a different combination of truth values for the atomic propositions $p, q, r, \ldots$.

This diagram can be made to look simpler, as in Figure 4.7, by leaving out the truth values for p and q – they are obvious – and by shading in those regions for which $p \wedge q$ is true (rather than explicitly stating the various truth values).

Examples 4.29 Represent each of the following in an appropriately shaded Venn diagram:

1. $p \vee q$
2. $p \Rightarrow q$
3. $\neg p$
4. $\neg p \vee q$

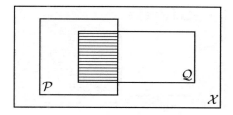

Figure 4.7 *Shaded area corresponds to $p \wedge q = T$.*

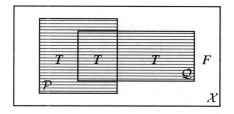

Figure 4.8 *Shaded area represents $p \vee q$.*

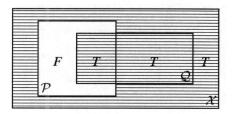

Figure 4.9 *Shaded area represents $p \Rightarrow q$.*

Solution 4.30

1. $p \vee q$ is represented in Figure 4.8.
2. $p \Rightarrow q$ is represented in Figure 4.9.
3. $\neg\, p$ is represented in Figure 4.10.
4. $\neg\, p \vee q$ is represented as truth values in Figure 4.11 and as the corresponding shaded area in Figure 4.12.

4.3.2   Logical connectives and set operations

From the definition of set comprehension, the set of points $\mathcal{P}$ for which $p(x)$ is true is given by

$$\mathcal{P} = \{\, x : \in \mathcal{X} \mid p(x)\,\}$$

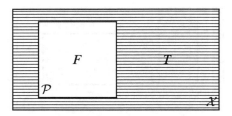

Figure 4.10 *Shaded area represents* $\neg\, p$.

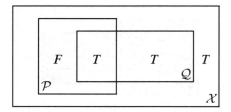

Figure 4.11 *Truth values for* $\neg\, p \vee q$.

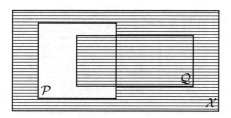

Figure 4.12 *Shaded area represents* $\neg\, p \vee q$.

Similarly for $\mathcal{Q}$:

$$\mathcal{Q} = \{\, x :\in \mathcal{X} \mid q(x) \,\}$$

In a like manner we can write the set of points for which $p(x) \wedge q(x)$ is T as

$$\{\, x :\in \mathcal{X} \mid p(x) \wedge q(x) \,\}$$

Now from the Venn diagram, it can be seen that this last set of points must be the overlap of $\mathcal{P}$ and $\mathcal{Q}$ – that is, the set of points which lie in both $\mathcal{P}$ and $\mathcal{Q}$. So by definition of set intersection, this set is

$$\mathcal{P} \cap \mathcal{Q}$$

or using the definitions of $\mathcal{P}$ and $\mathcal{Q}$

$$\{\, x :\in \mathcal{X} \mid p(x) \,\} \cap \{\, x :\in \mathcal{X} \mid q(x) \,\}$$

Hence we have the result

$$\{ x :\in \mathcal{X} \mid p(x) \land q(x) \} = \{ x :\in \mathcal{X} \mid p(x) \} \cap \{ x :\in \mathcal{X} \mid q(x) \}$$

Although we have obtained this result for the set of points $\mathcal{X}$ and for predicates to express a point lying inside a region, the result is true for *any* set $\mathcal{X}$ and predicates $p(x)$.

Examples 4.31 Find the corresponding relationship for $\lor$.

Solution 4.32 In this case $\lor$ corresponds to the *union* of the two sets in the Venn diagram.

$$\{ x :\in \mathcal{X} \mid p(x) \lor q(x) \} = \{ x :\in \mathcal{X} \mid p(x) \} \cup \{ x :\in \mathcal{X} \mid q(x) \}$$

Some specific examples with numbers rather than points will reinforce these results. For example $\{ x :\in \mathbb{N} \mid x > 3 \land x < 6 \} = \{4,5\}$. Now we know that $\{ x :\in \mathbb{N} \mid x > 3 \} = \{4,5,6,7,8,\ldots\}$ and $\{ x :\in \mathbb{N} \mid x < 6 \} = \{0,1,2,3,4,5\}$. Hence we can calculate $\{ x :\in \mathbb{N} \mid x > 3 \} \cap \{ x :\in \mathbb{N} \mid x < 6 \}$ as

$$\{4,5,6,7,8,\ldots\} \cap \{0,1,2,3,4,5\} = \{4,5\}$$

from which we can conclude that

$$\{ x :\in \mathbb{N} \mid x > 3 \land x < 6 \} = \{ x :\in \mathbb{N} \mid x > 3 \} \cap \{ x :\in \mathbb{N} \mid x < 6 \}$$

as expected.

Fact 4.33 If $p(x)$ and $q(x)$ are two predicates with free variable x whose values can range over $\mathcal{X}$, then

$$\{ x :\in \mathcal{X} \mid p(x) \land q(x) \} = \{ x :\in \mathcal{X} \mid p(x) \} \cap \{ x :\in \mathcal{X} \mid q(x) \}$$

The logical connective $\land$ corresponds to the set operation $\cap$.

Fact 4.34 If $p(x)$ and $q(x)$ are two predicates with free variable x whose values can range over a type T, then

$$\{ x :\in T \mid p(x) \lor q(x) \} = \{ x :\in T \mid p(x) \} \cup \{ x :\in T \mid q(x) \}$$

The logical connective $\lor$ corresponds to the set operation $\cup$.

Examples 4.35 Write each of the following set comprehensions as the union or intersection of two simpler set comprehensions:

1. $\{ x :\in \mathbb{Z} \mid x \in \mathbb{N} \land x < 8 \}$
2. $\{ x :\in \mathbb{Z} \mid x \in \mathbb{N} \lor x < 8 \}$
3. $\{ x :\in \mathbb{Z} \mid x \in \mathbb{N} \Rightarrow x < 8 \}$

4. $\{\, x :\in \mathbb{Z} \mid \neg\,(x \in \mathbb{N} \Rightarrow x < 8)\,\}$

Solution 4.36

1. $\{\, x :\in \mathbb{Z} \mid x \in \mathbb{N} \wedge x < 8\,\} = \{\, x :\in \mathbb{Z} \mid x \in \mathbb{N}\,\} \cap \{\, x :\in \mathbb{Z} \mid x < 8\,\}$
2. $\{\, x :\in \mathbb{Z} \mid x \in \mathbb{N} \vee x < 8\,\} = \{\, x :\in \mathbb{Z} \mid x \in \mathbb{N}\,\} \cup \{\, x :\in \mathbb{Z} \mid x < 8\,\}$
3. Since $p \Rightarrow q \equiv \neg\, p \vee q$ it follows that

$$\{\, x :\in \mathbb{Z} \mid x \in \mathbb{N} \Rightarrow x < 8\,\} = \{\, x :\in \mathbb{Z} \mid \neg\, x \in \mathbb{N} \vee x < 8\,\}$$

$$= \{\, x :\in \mathbb{Z} \mid \neg\, x \in \mathbb{N}\,\} \cup \{\, x :\in \mathbb{Z} \mid x < 8\,\}$$

4. Since $\neg\,(p \Rightarrow q) \equiv p \wedge \neg\, q$ it follows that

$$\{\, x :\in \mathbb{Z} \mid \neg\,(x \in \mathbb{N} \Rightarrow x < 8)\,\} = \{\, x :\in \mathbb{Z} \mid x \in \mathbb{N} \wedge \neg\, x < 8\,\}$$

$$= \{\, x :\in \mathbb{Z} \mid x \in \mathbb{N}\,\} \cap \{\, x :\in \mathbb{Z} \mid \neg\, x < 8\,\}$$

In fact this expression can be further simplified since

$$\{\, x :\in \mathbb{Z} \mid x \in \mathbb{N}\,\} \cap \{\, x :\in \mathbb{Z} \mid \neg\, x < 8\,\} = \mathbb{N} \cap \{\, x :\in \mathbb{Z} \mid x \geqslant 8\,\}$$

$$= \{\, x :\in \mathbb{Z} \mid x \geqslant 8\,\}$$

Examples 4.37 Write each of the following as a single set comprehension:

1. $\{\, x :\in \mathbb{Z} \mid x > 3\,\} \cap \{\, x :\in \mathbb{Z} \mid x < 7\,\}$
2. $\{\, y :\in \mathbb{Z} \mid y > 3\,\} \cup \{\, x :\in \mathbb{Z} \mid x < 9\,\}$
3. $\{\, x :\in \mathbb{Z} \mid x \leqslant 0\,\} \cup \{\, x :\in \mathbb{Z} \mid x > 3\,\}$

Solution 4.38

1. $\{\, x :\in \mathbb{Z} \mid x > 3\,\} \cap \{\, x :\in \mathbb{Z} \mid x < 7\,\} = \{\, x :\in \mathbb{Z} \mid x > 3 \wedge x < 7\,\}$.
   Conventionally the compound predicate $x > 3 \wedge x < 7$ would be written as $3 < x < 7$ so the set comprehension would be written as $\{\, x :\in \mathbb{Z} \mid 3 < x < 7\,\}$. We say that '$3 < x < 7$' is an *abbreviation* for '$x > 3 \wedge x < 7$'.
2. Since y is bound in the first set comprehension, it can be replaced by x to give $\{\, x :\in \mathbb{Z} \mid x > 3\,\} \cup \{\, x :\in \mathbb{Z} \mid x < 9\,\} = \{\, x :\in \mathbb{Z} \mid x > 3 \vee x < 9\,\}$. A little thought should convince you that $x > 3 \vee x < 9$ is true no matter what value is chosen for x, and so the filter predicate is redundant. The simplest form of the set comprehension is $\{\, x :\in \mathbb{Z}\,\}$, that is the answer is simply $\mathbb{Z}$. Note that $x > 3 \vee x < 9$ can be written as $\neg\, x \leqslant 3 \vee x < 9$ which is equivalent to $x \leqslant 3 \Rightarrow x < 9$; but we know that $x \leqslant 3 \Rightarrow x < 9$ is always true from the properties of numbers.
3. $\{\, x :\in \mathbb{Z} \mid x \leqslant 0\,\} \cup \{\, x :\in \mathbb{Z} \mid x > 3\,\} = \{\, x :\in \mathbb{Z} \mid x \leqslant 0 \vee x > 3\,\}$. In fact this could be written as $\{\, x :\in \mathbb{Z} \mid x > 0 \Rightarrow x > 3\,\}$ using the equivalence of $p \Rightarrow q$ and $\neg\, p \vee q$.

4.3.3 Metalanguage

We have seen that a region of the Venn diagram can represent a propositional form such as $p \wedge q$. Suppose that P and Q are propositional forms represented by the regions $\mathcal{P}$ and $\mathcal{Q}$. In general, the regions $\mathcal{P}$ and $\mathcal{Q}$ look something like Figure 4.13.

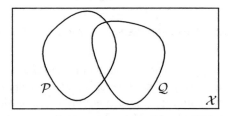

Figure 4.13 *General Venn diagram for $p(x)$ and $q(x)$.*

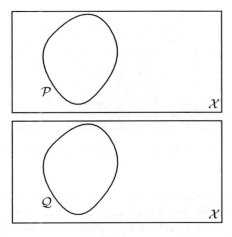

Figure 4.14 *Venn diagrams to represent equivalent propositional forms.*

However, it is sometimes the case that a special relationship exists between two propositional forms. Such a special relationship will be reflected in the relationship between the regions $\mathcal{P}$ and $\mathcal{Q}$ in the Venn diagram.

Equivalence We have already seen that the regions for $p \Rightarrow q$ and $\neg\, p \vee q$ are identical; this is because $p \Rightarrow q$ and $\neg\, p \vee q$ have the same truth tables – they are equivalent forms. In general, if the propositional forms P and Q are equivalent, then their corresponding regions are identical, as shown in Figure 4.14.

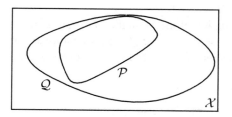

Figure 4.15 *Venn diagram to represent* $P \models Q$.

Logical implication Suppose that P logically implies Q, $P \models Q$, that is, whenever P is true Q is also true. Then the region $\mathcal{P}$ (corresponding to P) must lie wholly within the region $\mathcal{Q}$ (corresponding to Q), as shown in Figure 4.15. To illustrate this consider the Venn diagram for $p \Rightarrow q$ and $p \wedge q$, as shown in Figure 4.16. The region for $p \wedge q$ lies entirely within $p \Rightarrow q$, and so we can conclude that $p \Rightarrow q \models p \wedge q$.

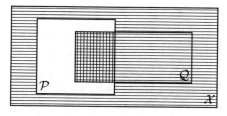

Figure 4.16 *Venn diagram for* $p \wedge q$ *(vertical shading) and* $p \Rightarrow q$ *(horizontal shading).*

Tautologies and contradictions can also be represented in a Venn diagram. If the propositional form P is a tautology then its truth value is always equal to T. Now recall that the region $\mathcal{P}$ represents those instances for which P is true. It follows that the region $\mathcal{P}$ must occupy the whole box, as shown in Figure 4.17. In fact the

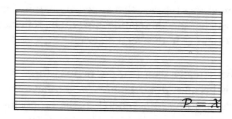

Figure 4.17 *Representation of a tautology* $\mathcal{P}$.

box itself can be regarded as representing the propositional form T.

If however P is a contradiction then there can be no shaded region corresponding to it; $\mathcal{P} = \varnothing$, as illustrated in Figure 4.18.

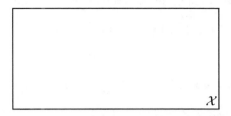

Figure 4.18 *Representation of a contradiction.*

An association between $\Rightarrow$ and $\models$
Suppose $P \models Q$, then we can construct the region corresponding to $P \Rightarrow Q$ as shown in Figure 4.19. Because of the special relationship between $\mathcal{P}$ and $\mathcal{Q}$, there

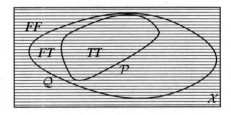

Figure 4.19 *Venn diagram for $P \Rightarrow Q$ when $P \models Q$.*

is no region inside the box for which P is T while Q is F; thus $P \Rightarrow Q$ corresponds to the whole box, that is $P \Rightarrow Q$ is a tautology: $\models P \Rightarrow Q$.

4.4   Quantifiers

By enclosing expression $x :\in \mathbb{N}_1 \mid x < 4 \bullet x^2$ in braces $\{\ldots\}$ we can form a single object, namely the set of squares of positive integers less than 4:

$$\{x :\in \mathbb{N}_1 \mid x < 4 \bullet x^2\}$$

By placing the *summation operator* $\sum$ in front, we can form another single object, $\sum x :\in \mathbb{N}_1 \mid x < 4 \bullet x^2$, whose value is in fact the sum of the squares of the positive integers less than 4. (The value is 14.)

In both these examples we have produced a single object by means of a *quantifier*: in the first example, the set quantifier $\{\ldots\}$; in the second example, the summation quantifier $\sum$. In each case, the variable x is no longer free; we say that it has been *bound* by the quantifier. Thus in either case we could replace the *bound variable x* by any other variable and still refer to the same object. For example,

$$\{\, x :\in \mathbb{N}_1 \mid x < 4 \bullet x^2 \,\} \equiv \{\, y :\in \mathbb{N}_1 \mid y < 4 \bullet y^2 \,\}$$

or

$$\sum x :\in \mathbb{N}_1 \mid x < 4 \bullet x^2 \equiv \sum y :\in \mathbb{N}_1 \mid y < 4 \bullet y^2$$

Examples 4.39 Evaluate $\sum x :\in \mathbb{N} \mid x < 3 \bullet x$.

Solution 4.40

$$\sum x :\in \mathbb{N} \mid x < 3 \bullet x = 0 + 1 + 2 = 3$$

Two important quantifiers used in logic are the *universal quantifier*, $\forall$, and the *existential quantifier*, $\exists$. In order to explain these two quantifiers, we shall first take a look at how a declaration and predicate can be associated with a family of propositions.

4.4.1   Families of propositions from predicates

We have already seen that a declaration together with a predicate can be used to generate the elements of a set; this is done by placing a vertical line $\mid$ ('such that') between the declaration and the predicate. For example $x :\in \mathbb{N}_1 \mid x < 4$ corresponds to a process which generates values of x equal to 1, 2 and 3; $\{\, x :\in \mathbb{N}_1 \mid x < 4 \,\}$ is the set of all these possible values, that is $\{1, 2, 3\}$. We have seen how this idea can be extended by the use of the spot notation; each value generated can be further processed, so that, for example, $x :\in \mathbb{N}_1 \mid x < 4 \bullet x^2$ generates values of 1, 2 and 3 and then inputs these values to a squaring procedure (x^2). The corresponding set of values $\{\, x :\in \mathbb{N}_1 \mid x < 4 \bullet x^2 \,\}$ is $\{1^2, 2^2, 3^2\}$, which is equal to $\{1, 4, 9\}$.

In fact the expression after the spot might itself be another predicate; for example we might have $x :\in \mathbb{N}_1 \mid x < 4 \bullet x^2 > 5$. This expression can be regarded as 'generating' a family of propositions with each of which we can associate a truth value.

Examples 4.41 For each of the following expressions, tabulate the possible values of the free variable x, the corresponding propositions and their truth values:

1. $x :\in \mathbb{N}_1 \mid x < 4 \bullet x^2 > 5$
2. $x :\in \mathbb{N} \mid x < 4 \bullet x^2 > 5$
3. $x :\in \mathbb{N} \mid x^2 < 5 \bullet x > 5$
4. $x :\in \mathbb{N} \mid x \in \{2, 3\} \bullet x^2 < 20$

Solution 4.42

1. $x :\in \mathbb{N}_1 \mid x < 4 \bullet x^2 > 5$. The possible values of x are $1, 2$ and 3.

x value	Proposition	Truth Value
1	$1^2 > 5$	F
2	$2^2 > 5$	F
3	$3^2 > 5$	T

2. $x :\in \mathbb{N} \mid x < 4 \bullet x^2 > 5$. The possible values of x are drawn from $\mathbb{N}$, which includes 0; the possible values of x are $0, 1, 2$ and 3.

x value	Proposition	Truth Value
0	$0^2 > 5$	F
1	$1^2 > 5$	F
2	$2^2 > 5$	F
3	$3^2 > 5$	T

3. $x :\in \mathbb{N} \mid x^2 < 5 \bullet x > 5$. Note that the filter predicate may involve a function of x. The possible values of x are $0, 1$ and 2.

x value	Proposition	Truth Value
0	$0 > 5$	F
1	$1 > 5$	F
2	$2 > 5$	F

Notice in this case that all the propositions are false.

4. $x :\in \mathbb{N} \mid x \in \{2, 3\} \bullet x^2 < 20$. In this case the possible values of x are just those in the set $\{2, 3\}$.

x value	Proposition	Truth Value
2	$2^2 < 20$	T
3	$3^2 < 20$	T

In this case all the propositions are true.

Note that in some cases we have an infinite family of propositions, so we would not be able to list them all.

4.4.2 Existential and universal quantifiers

Suppose we have a family of propositions generated by a predicate, for example:

$$x :\in \mathbb{N} \mid x < 3 \bullet x^2 > 3 \tag{4.1}$$

We might wish to make statements about the truth values of this family of propositions. For example we might assert that all the propositions are true, or that there is at least one instance of a true proposition. These assertions can be made by use of quantifiers.

Definition 4.43 The *Universal Quantifier*, ∀, to indicate that all the propositions are true.

Definition 4.44 The *Existential Quantifier*, ∃, to indicate that at least one proposition is true.

For example,

$$\forall x :\in \mathbb{N} \mid x < 3 \bullet x^2 > 3 \tag{4.2}$$

$$\exists x :\in \mathbb{N} \mid x < 3 \bullet x^2 > 3 \tag{4.3}$$

Note that in each case the result is a *proposition*; our assertion may itself be true or false. The intuitive interpretation of the proposition 4.2 is that *all* the separate propositions generated by the predicate $x^2 > 3$ are true. Similarly the interpretation of the proposition 4.3 is that *at least one* of the propositions generated by the predicate in $x^2 > 3$ is true.

In these particular cases the typed predicate generates a finite, and indeed small, number of propositions, so we can easily determine the truth of propositions 4.2 and 4.3 by producing a table of the propositions and their corresponding truth values.

x	$x^2 > 3$	Truth Value
0	$0^2 > 3$	F
1	$1^2 > 3$	F
2	$2^2 > 3$	T

From the table we can see that two propositions are false, but one is true. Thus it follows that $\forall x :\in \mathbb{N} \mid x < 3 \bullet x^2 > 3 = F$ but $\exists x :\in \mathbb{N} \mid x < 3 \bullet x^2 > 3 = T$.

Examples 4.45 Find the truth values of the following:

1. $\forall x :\in \mathbb{N}_1 \mid x < 4 \bullet x > 0$
2. $\exists x :\in \mathbb{N}_1 \mid x < 4 \bullet x > 0$
3. $\exists x :\in \mathbb{N} \mid x < 3 \bullet \neg\, x^2 > 3$
4. $\neg\, \exists x :\in \mathbb{N} \mid x < 3 \bullet x^2 > 3$

Solution 4.46

1. $\forall x :\in \mathbb{N}_1 \mid x < 4 \bullet x > 0 = T$
2. $\exists x :\in \mathbb{N}_1 \mid x < 4 \bullet x > 0 = T$

x	$x > 0$	Truth Value
1	$1 > 0$	T
2	$2 > 0$	T
3	$3 > 0$	T

3. $\exists x :\in \mathbb{N} \mid x < 3 \bullet \neg\, x^2 > 3 = T$

4. $\neg \ \exists x :\in \mathbb{N} \mid x < 3 \bullet x^2 > 3 = F$

x	$x^2 > 3$	TruthValue	$\neg \ x^2 > 3$	TruthValue
0	$0^2 > 3$	F	$\neg \ 0^2 > 3$	T
1	$1^2 > 3$	F	$\neg \ 1^2 > 3$	T
2	$2^2 > 3$	T	$\neg \ 2^2 > 3$	F

Note that the last two answers are different; although the two expressions *look* similar, they are different.

4.4.3 Predicates with more than one free variable

So far we have been using the universal and existential quantifiers over a one place predicate, the quantifier binding the single variable. However we often apply quantifiers to predicates with more than one variable, for example in

$\forall x :\in \mathbb{Z} \bullet y < x$

the variable x is bound by $\forall$, but y is free. We therefore have a predicate in y. We can use this predicate in a typed expression of the form

$y :\in T \bullet p(y)$

where T is the set of possible y values and $p(y)$ is the predicate in y; in this case

$p(y) = \forall x :\in \mathbb{Z} \bullet y < x$

so that the typed expression is

$y :\in \mathbb{Z} \bullet (\forall x :\in \mathbb{Z} \bullet y < x)$

assuming that the type of y is $\mathbb{Z}$. It is, of course, possible to bind the variable y with a quantifier, for example

$\exists y :\in \mathbb{Z} \bullet (\forall x :\in \mathbb{Z} \bullet y < x)$

$\forall y :\in \mathbb{Z} \bullet (\forall x :\in \mathbb{Z} \bullet y < x)$

Such expressions can be evaluated for small sets of x and y by tabulation.

Examples 4.47 Evaluate:

1. $\exists y :\in \mathbb{N} \mid y < 2 \bullet (\forall x :\in \mathbb{N} \mid x < 3 \bullet y < x)$
2. $\forall y :\in \mathbb{N} \mid y < 2 \bullet (\exists x :\in \mathbb{N} \mid x < 3 \bullet y < x)$

Solution 4.48 In each case the possible values of x are 0, 1, 2 and of y are 0, 1. Both $\forall x :\in \mathbb{N} \mid x < 3 \bullet y < x$ and $\exists x :\in \mathbb{N} \mid x < 3 \bullet y < x$ need to be evaluated for each of the y values:

- If y is set to 0 then we have

x	$y < x$	TruthValue
0	$0 < 0$	F
1	$0 < 1$	T
2	$0 < 2$	T

In this case

$$\forall x :\in \mathbb{N} \mid x < 3 \bullet y < x = F$$

while

$$\exists x :\in \mathbb{N} \mid x < 3 \bullet y < x = T$$

- If y is set to 1 then we have

x	$y < x$	TruthValue
0	$1 < 0$	F
1	$1 < 1$	F
2	$1 < 2$	T

Again we find that

$$\forall x :\in \mathbb{N} \mid x < 3 \bullet y < x = F$$

while

$$\exists x :\in \mathbb{N} \mid x < 3 \bullet y < x = T$$

These results enable us to conclude that

1. $\forall x :\in \mathbb{N} \mid x < 3 \bullet y < x$ is not true for either of the y values so

$$\exists y :\in \mathbb{N} \mid y < 2 \bullet (\forall x :\in \mathbb{N} \mid x < 3 \bullet y < x) = F$$

2. $\exists x :\in \mathbb{Z} \mid x < 3 \bullet y < x$ is true for both values of y so

$$\forall y :\in \mathbb{N} \mid y < 2 \bullet (\exists x :\in \mathbb{N} \mid x < 3 \bullet y < x) = T$$

Fact 4.49 If $P(x, y)$ is a predicate with free variables $x :\in X$ and $y :\in Y$ then it may be that $\forall y :\in Y \bullet \exists x :\in X \bullet P(x, y)$ and $\exists y :\in Y \bullet \forall x :\in X \bullet P(x, y)$ have different truth values. Indeed the two expressions are saying different things!

Another important result (see exercise at end of chapter for examples) is:

Fact 4.50 If $P(x, y)$ is a predicate with free variables $x :\in X$ and $y :\in Y$ then in general $\forall y :\in Y \bullet \exists x :\in X \bullet P(x, y)$ and $\exists x :\in X \bullet \forall y :\in Y \bullet P(x, y)$ are different.

Clearly great care needs to be used with quantifiers of more than one variable, to ensure that the quantifiers are placed in the right order. There are nevertheless certain special cases in which changing the order leaves the truth value unchanged.

Fact 4.51 If $P(x, y)$ is a predicate with free variables $x :\in X$ and $y :\in Y$ then both $\forall y :\in Y \bullet \forall x :\in X \bullet P(x, y)$ and $\forall x :\in X \bullet \forall y :\in Y \bullet P(x, y)$ always have the same truth value no matter what is chosen for X, Y and $P(x, y)$. We say that the two expressions are *equivalent* and represent this equivalence by the symbol $\equiv$ as follows:

$$\forall y :\in Y \bullet \forall x :\in X \bullet P(x, y) \equiv \forall x :\in X \bullet \forall y :\in Y \bullet P(x, y)$$

Notation 4.52 Two universal quantifications are often combined into a much neater expression $\forall x :\in X, y :\in Y \bullet P(x, y)$ since the two possible 'long forms' are equivalent.

A similar result holds for the existential quantifier.

Fact 4.53

$$\exists y :\in Y \bullet \exists x :\in X \bullet P(x, y) \equiv \exists x :\in X \bullet \exists y :\in Y \bullet P(x, y)$$

Again we can represent either expression more neatly as

Notation 4.54

$$\exists x :\in X, y :\in Y \bullet P(x, y)$$

Notation 4.55 Where expression forms do not always have the same truth values, we say that they are not equivalent and denote this by $\not\equiv$. For example

$$\forall y :\in Y \bullet \exists x :\in X \bullet P(x, y) \not\equiv \exists x :\in X \bullet \forall y :\in Y \bullet P(x, y)$$

4.4.4 Set comprehension

In a set comprehension of the form $y :\in T \mid P(y)$ the predicate $P(y)$ has free variable y, but may have one or more bound variables. For example in

$$\{ y :\in \mathbb{Z} \mid (\forall x :\in \mathbb{N} \bullet y < x) \}$$

the predicate has the variable x bound by the universal quantifier, $\forall$.

Examples 4.56 Find set enumerations for each of the following, and where possible obtain a simpler set comprehension:

1. $\{ y :\in \mathbb{N} \mid y < 3 \wedge (\exists x :\in \mathbb{N} \mid x < 2 \bullet y = 2 * x) \}$
2. $\{ y :\in \mathbb{N} \mid y < 3 \wedge (\exists x :\in \mathbb{N} \mid x < 4 \bullet x = 2 * y) \}$
3. $\{ y :\in \mathbb{P}\mathbb{N} \mid (\forall x :\in \mathbb{N} \mid x < 2 \bullet x \in y) \}$

Solution 4.57

1. The possible values of x are 0 and 1; furthermore $y < 3$ is true only for $y = 0, 1, 2$ (for $y :\in \mathbb{N}$) so we only need to consider these values in any predicate of the form $y < 3 \wedge P(x, y)$.

   - Now when $y = 0$ we can find a value of x, namely 1, for which $y = 2 * x$ is true, so $\exists x :\in \mathbb{N} \mid x < 2 \bullet y = 2 * x$ is true for $y = 0$; 0 is a member of our set.

   - Similarly when $y = 2$ we can find a value of x, namely 1, for which $y = 2 * x$ is true, so again $\exists x :\in \mathbb{N} \mid x < 2 \bullet y = 2 * x$ is true for $y = 2$; 2 is a member of our set.

   - But when $y = 1$, neither of the possible values of x (that is 0 and 1) make $y = 2 * x$ true, so $\exists x :\in \mathbb{N} \mid x < 2 \bullet y = 2 * x$ is false for $y = 1$; 1 is not a member of our set.

   The required set enumeration is $\{0, 2\}$. This is the set of *even* values of y from $\mathbb{N}$ which are less than 3. Note that $\exists x :\in \mathbb{N} \bullet y = 2 * x$ can be used to express the fact that y is exactly divisible by 2, and so a simpler set comprehension is

   $$\{\, y :\in \mathbb{N} \mid y < 3 \wedge (\exists x :\in \mathbb{N} \bullet y = 2 * x) \,\}$$

   An even simpler set comprehension is $\{\, x :\in \mathbb{N} \mid x < 2 \bullet 2 * x \,\}$.

2. Each element y of $\{\, y :\in \mathbb{N} \mid y < 3 \wedge (\exists x :\in \mathbb{N} \mid x < 4 \bullet x = 2 * y) \,\}$ is a natural number less than 3, $y < 3$, such that twice its value, $2 * y$, is a natural number x less than 4. The only possible values are 0 and 1 so the set enumeration is $\{0, 1\}$. A much simpler set comprehension is $\{\, x :\in \mathbb{N} \mid x < 2 \,\}$.

3. Each element y of $\{\, y :\in \mathbb{P}\mathbb{N} \mid (\forall x :\in \mathbb{N} \mid x < 2 \bullet x \in y) \,\}$ is itself a set of natural numbers, $y :\in \mathbb{P}\mathbb{N}$, such that every number x less than 2 is an element of the set y; that is both 0 and 1 must be elements of y. This is the set of all subsets of $\mathbb{N}$ which contain both 0 and 1. It is an infinite set, and so cannot be written as a set enumeration. A simpler set comprehension is $\{\, y :\in \mathbb{P}\mathbb{N} \bullet y \cup \{0, 1\} \,\}$.

4.5   Case study: project teams

Predicate logic is very useful in defining properties. For example, in section 2.4 we considered project teams drawn from a workforce. We considered a particular, small example where we had a set *Workforce* equal to $\{Elma, Rajesh, Mary, Carlos, Mike\}$ and from which we took four project teams: *project_A*, *project_B*, *project_C* and *project_D*. Now suppose we want to impose constraints on the membership of project teams; for example, that each team should include at least one experienced person. We can do this by defining a subset *experienced* of *Workforce*, and then insisting that there is at least one member of each project team that is a member of *experienced*. How can we express this desired property using predicate logic?

One approach to getting an answer is to suppose we have a project team *project* from which we can choose a *person*. We want to know whether or not this *person* is experienced, that is whether

$$person \in experienced$$

is true or false. In fact we want to do this for every possible value of *person* chosen from the set *experienced* to generate a family of propositions (note that *person* is the free variable).

$$person :\in project \bullet person \in experienced \tag{4.4}$$

Suppose for example that

$$experienced = \{Elma, Rajesh\}$$

and that

$$project = \{Elma, Carlos\}$$

then expression (4.4) would generate the propositions

$$Elma \in experienced$$

which is true and

$$Carlos \in experienced$$

which is false.

Now we want at least one of these propositions to be true; we can express this by the existential quantifier

$$\exists person :\in project \bullet person \in experienced \tag{4.5}$$

In the simple example just given, this proposition is indeed true. We want this proposition to be true for all acceptable project teams.

We begin by considering *project* to be a free variable which can be chosen from the set of all subsets of *Workforce* (that is, the power set of *Workforce*).

$$project :\in \mathbb{P}\, Workforce$$

Now consider the family of propositions generated by

$$project :\in \mathbb{P}\, Workforce \bullet (\exists person :\in project \bullet person \in experienced)$$

In general, not all the propositions generated by this expression will be true. We therefore need to restrict our choice of *project* to a subset of $\mathbb{P}$ *Workforce*, that is we define a new set

$$acceptable :\in \mathbb{P}\,\mathbb{P}\ Workforce$$

and consider the set of propositions generated by

$$project :\in acceptable \bullet (\exists\, person :\in project \bullet person \in experienced) \quad (4.6)$$

We require that the set *acceptable* has the property that all the propositions generated by (4.6) are true. We can represent this by means of the universal quantifier.

$$\forall\, project :\in acceptable \bullet (\exists\, person :\in project \bullet person{\in}experienced) \quad (4.7)$$

This captures our required property for any project team to contain at least one experienced member. It is important to note that expression (4.7) can be applied to any *Workforce* and any *experienced* set of people.

4.6   Exercise

1. Which of the following are elements of $\{\,x :\in \mathbb{Z} \mid x < 4\,\}$?
   (a) 1
   (b) 7
   (c) -1

2. Which of the following are elements of $\{\,x :\in \mathbb{P}\,\mathbb{N} \mid \#x < 2\,\}$?
   (a) $\varnothing$
   (b) 1
   (c) $\{1\}$
   (d) $\{1001\}$
   (e) $\{3.14152\}$
   (f) $\{0,1\}$

3. Which of the following are elements of $\{\,x :\in \mathbb{N} \times \mathbb{Z} \mid x.1 < 4 \land x.2 < 5\,\}$?
   (a) 2
   (b) $(1,2)$
   (c) $\{1,2\}$
   (d) $\{(1,2)\}$
   (e) $(4,2)$

(f) $(3, -1)$

(g) $(-1, 3)$

4. Find the set enumerations for

(a) $\{ x :\in \mathbb{N}_1 \mid x < 5 \}$

(b) $\{ y :\in \mathbb{N} \mid y^2 < 20 \}$

(c) $\{ z :\in \mathbb{Z} \mid z > 4 \wedge z < 7 \}$

(d) $\{ x :\in \mathbb{N} \mid x < 4 \bullet 3 * x \}$

(e) $\{ x :\in \mathbb{Z} \mid x > -3 \wedge x < 3 \bullet x^2 \}$

(f) $\{ x :\in \mathbb{N}_1 \mid x < 4 \bullet \{x\} \}$

(g) $\{ x :\in \mathbb{Z} \mid x > -3 \wedge x < 3 \bullet (x, x^2) \}$

(h) $\{ x, y :\in \mathbb{N} \mid x < 4 \wedge y < 1 \bullet (x, y) \}$

(i) $\{ x, y :\in \mathbb{N} \mid x < 4 \wedge y < 1 \bullet (x, x + y) \}$

(j) $\{ x, y :\in \mathbb{Z} \mid -2 < x < 3 \wedge 0 < y < x \bullet (x, y) \}$

5. In each of the following expressions, which are the free variables and which are bound?

(a) $y^2 > 5$

(b) $\{ x :\in \mathbb{N}_1 \mid x + 2 = 1 \}$

(c) $x \in \{ y :\in \mathbb{Z} \mid y^2 < 0 \}$

6. Express the following sets using set comprehension, if necessary using the $\bullet$ extension.

(a) $\{0, 4, 8, 12\}$

(b) $\{\{0\}, \{1\}, \{2\}\}$

(c) $\{\{0\}, \{3\}, \{6\}\}$

(d) $\{(0, 0), (1, 2), (2, 4), (3, 6), (4, 8)\}$

(e) $\{(0, 0), (2, 2), (4, 4), (6, 6), (8, 8)\}$

7. Write each of the following set comprehensions as the union or intersection of two simpler set comprehensions:

(a) $\{ x :\in \mathbb{Z} \mid x > 10 \vee x < 8 \}$

(b) $\{ x :\in \mathbb{Z} \mid 6 < x < 8 \}$

(c) $\{ x :\in \mathbb{Z} \mid x \in \mathbb{N} \Rightarrow x < 8 \}$

(d) $\{ x :\in \mathbb{Z} \mid \neg\, (x \in \mathbb{N} \Rightarrow x < 8) \}$

8. Write each of the following as a single set comprehension:

(a) $\{ x :\in \mathbb{Z} \mid x > 6 \} \cap \{ x :\in \mathbb{Z} \mid x < 7 \}$

(b) $\{ y :\in \mathbb{Z} \mid y > 6 \} \cup \{ x :\in \mathbb{Z} \mid x < 9 \}$

(c) $\{x :\in \mathbb{Z} \mid x \leqslant 0\} \cup \{x :\in \mathbb{Z} \mid x > 6\}$

9. Evaluate $\sum x :\in \mathbb{N} \mid x < 3 \bullet x^2$.
10. For each of the following expressions, tabulate the possible values of the free variables, the corresponding propositions and their truth values; for the last three examples, there are infinitely many values of the free variables, so take just a few sample values, some of which give a true proposition and some of which give a false proposition.

    (a) $x :\in \mathbb{N} \mid x < 5 \bullet x^2 > 4$

    (b) $y :\in \mathbb{N} \mid y < 5 \bullet y < 20$

    (c) $z :\in \mathbb{N}_1; \; x :\in \mathbb{N} \mid z < 4 \wedge x < 2 \bullet (z^2 > 4 \Rightarrow x * z > 4)$

    (d) $w, y :\in \mathbb{N} \mid y < 3 \wedge w < y \bullet (w^2 > 4 \Leftrightarrow \neg y^2 > 4)$

    (e) $y :\in \mathbb{PN} \bullet \#y = 5$

    (f) $z :\in \mathbb{PN} \bullet \#z = 5$

    (g) $x :\in \mathbb{N}, y :\in \mathbb{PN} \bullet \#y = x)$

11. Determine the truth value for each of the following statements:

    (a) $\forall x :\in \mathbb{N} \mid x < 5 \bullet x^2 > 4$

    (b) $\exists x :\in \mathbb{N} \mid x < 5 \bullet x^2 > 4$

    (c) $\forall y :\in \mathbb{N} \mid y < 5 \bullet y < 20$

    (d) $\exists y :\in \mathbb{N} \mid y < 5 \bullet y < 20$

    (e) $\forall y :\in \mathbb{PN} \bullet \#y = 5$

    (f) $\exists z :\in \mathbb{PN} \bullet \#z = 5$

    (g) $\forall z :\in \mathbb{N}_1 \mid z < 4 \bullet (z^2 > 4 \Rightarrow \exists x :\in \mathbb{N} \mid z < x \bullet x * z > 4)$

    (h) $\neg \forall w :\in \mathbb{N} \mid w < 3 \bullet (w^2 > 4 \Leftrightarrow \exists y :\in \mathbb{N} \mid w < y \bullet \neg y^2 > 4)$

    (i) $\neg \forall w :\in \mathbb{N} \mid w < 3 \bullet (w < 20 \Leftrightarrow \exists y :\in \mathbb{N} \mid w < y \bullet \neg y < 20)$

    (j) $\forall y :\in \mathbb{PN} \bullet (\exists x :\in \mathbb{N} \bullet \#y = x)$

    (k) $\exists z :\in \mathbb{PN} \bullet (\forall x :\in \mathbb{N} \bullet \#z = x)$

12. Find set enumerations for each of the following, and where possible obtain a simpler set comprehension.

    (a) $\{y :\in \mathbb{N}_1 \mid y < 7 \wedge (\exists x :\in \mathbb{Z} \mid x < 3 \bullet y = 3 * x)\}$

    (b) $\{y :\in \mathbb{N} \mid y < 7 \wedge (\exists x :\in \mathbb{Z} \mid x < 4 \bullet x = 3 * y)\}$

    (c) $\{y :\in \mathbb{N}_1 \mid y \leqslant 4 \wedge (\exists x :\in \mathbb{Z} \mid x < 4 \bullet 6 = y * x)\}$

    (d) $\{y :\in \mathbb{N}_1 \mid y < 7 \wedge (\forall x :\in \mathbb{Z} \mid x < 4 \bullet (\exists z :\in \mathbb{Z} \bullet y = z * x))\}$

    (e) $\{y :\in \mathbb{PN} \mid (\exists x :\in \mathbb{N} \mid x < 3 \bullet x \in y)\}$

13. Suppose we have (as given in section 2.4)

$$Workforce = \{Elma, Rajesh, Mary, Carlos, Mike\}$$
$$project\_A = \{Rajesh, Mike, Carlos\}$$
$$project\_B = \{Mary, Rajesh, Mike, Elma\}$$
$$project\_C = \{Elma, Carlos\}$$
$$project\_D = \{Rajesh, Mary, Mike, Carlos\}$$

Mike has been given the task of setting up a new project team (*new_project*). Write down a proposition to capture each of the following requirements:

(a) *Mike* must be a member of *new_project*.

(b) *new_project* must have at least two members.

(c) *Elma* will only be a member of *new_project* if *Rajesh* is.

(d) At least one member of *new_project* must be a member of *project_A*.

(e) Every member of *new_project* must be a member of *project_A*.

(f) Every member of *project_C* must be a member of *new_project*.

(g) Either *new_project* has at least one member of *project_C* or it has every member of *project_D*.

(h) Every member of *new_project* is a member of both *project_A* and *project_B*.

(i) Every member of *new_project* is a member of either *project_A* or *project_B*.

CHAPTER 5

Relations

5.1   What is a relation?

Much of computing, and indeed of human activity, is concerned with handling and processing information. In this chapter, we shall be looking at how sets and logic can be used to model information and information processing through the concept of *relations*.

Consider the following data.

Mike is taller than Carlos.

Sarah is a parent of Ben.

Daniel is a parent of Ben.

One approach to modelling facts such as these, is to regard the sentences as propositions to each of which we assign the value T. This approach is fine provided we are concerned with isolated facts or data. But there is much more to the information than isolated facts; we need to be able to organize and process data to give useful information, or to answer queries such as

Who is taller than Carlos?

Of whom is Sarah a parent?

One way of organizing facts is to use the concept of a relation – a relation corresponds to a *property* such as 'is taller than' or 'is a parent of'. Thus we can stipulate that the property 'is taller than' applies to Mike and Carlos:

is_taller(Mike, Carlos)

An alternative way of expressing the relationship between Mike and Carlos is to place *is_taller* in the middle:

Mike *is_taller* Carlos

This way of writing a relation is referred to as *infix* notation. In general if *x* is related to *y* by *R*, then using infix notation we can write *x R y*.

Often we want to use a relation to obtain information. For example we may want to know 'Who is taller than Carlos?'. This question is equivalent to finding the values of *x* which fit the pattern:

is_taller(*x*, Carlos)

or using infix notation

x is_taller Carlos

With our *database* the only possible value of *x* is 'Mike'.

5.1.1   The order of a relation

A relation such as *is_taller* is said to be a *binary* relation, or relation of order two, since it requires two objects (Mike and Carlos, say) to make a complete statement. A relation such as *has_blue_eyes* is a *unary*, or order one, relation, since only one object is required to complete the sense; for example 'Elzelina has blue eyes'. Similarly we can define *ternary* relations (order three), *quaternary* relations (order four) and so on. Generally we shall consider just binary relations in this book; most concepts applicable to binary relations can be generalized to relations of higher order.

5.2   Operations on relations

The study of relations will itself necessitate the study of operations on relations. Some of these will be considered briefly here; later, mathematical models will be developed which will make the meanings of these operations much clearer.

5.2.1   Inverse or converse relations

Any (binary) relation has an *inverse* relation, also known as a *converse* relation. For example, *is_the_parent_of* is the inverse of *is_the_child_of*. In a sense, the inverse relation reverses the original. So if

- Mary *is_the_parent_of* Carlos
- Mike *is_the_parent_of* Carlos

then we can say

- Carlos *is_the_child_of* Mary
- Carlos *is_the_child_of* Mike

5.2.2 Composition of relations

Often a single relation can be thought of as being built up from two or more relations. For example *is_the_aunt_of* can be thought of as combining *is_the_sister_of* with *is_the_parent_of*. We say that *is_the_aunt_of* is equal to the *composition* of *is_the_sister_of* with *is_the_parent_of*. Suppose that

- Lisa *is_the_sister_of* Mike and
- Mike *is_the_parent_of* Rajesh

then it follows that Lisa *is_the_sister_of* a person who *is_the_parent_of* Rajesh, that is Lisa *is_the_aunt_of* Rajesh.

5.2.3 Relation override

Information is rarely static; relations will necessarily change. One way in which we *could* handle the changes in a relation would be to completely restate the relation every time there is a change. This would be hopelessly inefficient in all but the simplest cases. For example a company might keep records of the towns covered by each member of its sales team. If there is a reallocation of towns over part of the sales force, then a relation can be defined which covers only those people affected; this relation can then be used to *override*, and hence update, the total relation – information relating to those people unaffected will be unchanged by the override.

This concept may seem vague and difficult to understand. One of the reasons for introducing models of relations is to give clearer definitions of concepts such as override. At this stage, readers should recognize that their confusion is a sign of the need for discrete mathematics rather than a lack of their own intelligence!

5.3 Modelling relations

The concept of a relation finds use in areas such as relational databases and logic programming (the 'Prolog' language, for example). Unfortunately we have not yet given a clear, precise definition of what a relation is; without such an understanding, relational databases and logic programming are likely to be bewildering, especially to the beginner. Models of the concept of a relation can be built using the sets and logic we have considered in this book. In particular there are two common ways of viewing a relation:

- as a predicate;
- or as a set of ordered *tuplets*.

These two approaches are very closely related and often both are useful in the same context; for example, a Prolog database can be considered as sets of ordered pairs.

We shall first look *briefly* at the use of predicates to model relations. Most of what is needed here has already been covered. We shall then look at the use of sets of ordered pairs to model relations in greater detail; modelling relations with sets is a very useful approach and warrants detailed study.

5.3.1   Relations as predicates

The assertion 'Mike is taller than Carlos' can be regarded as the assignment of the truth value T to the proposition

Mike *is_taller* Carlos

This proposition in turn can be regarded as an instance of the two place (or *binary*) predicate

x *is_taller* y

and is obtained by substituting Mike for x and Carlos for y.

We can define new relations in terms of existing ones. For example we may define x *is_the_offspring_of* y to be an *abbreviation* for

y *is_the_parent_of* x

Connectives and quantifiers can also be used to build more complex predicates and to define new relations. For example, x *is_the_aunt_of* y can be defined as

$\exists z \bullet x$ *is_the_sister_of* $z \wedge z$ *is_the_parent_of* y

Metalanguage and corresponding metasymbols can also be used to express relations between predicates – we have in effect relations between relations! For example:

x *is_the_offspring_of* $y \models y$ *is_the_parent_of* x

since whenever the first relation holds, the second relation also holds.

5.4   Modelling relations with product sets

5.4.1   Binary relations

In arriving at a model we note that the order in which the related objects are stated is important. The statements 'Mike is taller than Carlos' and 'Carlos is taller than Mike' do not mean the same thing. Thus we are led to the idea of modelling a binary relation by a set of ordered pairs, with each ordered pair corresponding to a true proposition. The relation *is_taller* might for example be modelled by the set

$\{(Mike, Carlos), (Nasrin, Mike), (Nasrin, Carlos)\}$

from which we can see that Nasrin is taller than both Mike and Carlos, while Mike is also taller than Carlos.

In this case we think of the relation *is_taller* as referring to the set *People* which could, for example, be equal to {*Nasrin, Carlos, Mike*}; the corresponding ordered pair model is in fact a subset of *People* × *People*. Any binary relation on a set *A* will be modelled by a subset of *A* × *A*.

In general, however, a binary relation exists between the elements of one set and those of another set. For example we might have the relation *owns* between *People* and *Transport*. The relation *owns* could then be modelled by a subset of *People* × *Transport*. The first set (in this case *People*) is commonly called the *source* and the second set (in this case *Transport*) the *target* or *codomain*.

Suppose, for example, that the source set is given by

$$People = \{Mike, Carlos, Nasrin\}$$

and the target set is given by

$$Transport = \{Car, Bike, Horse, Plane\}$$

Then the relation *owns* will be modelled by a set of ordered pairs, for example:

$$\{(Nasrin, Bike), (Nasrin, Car), (Carlos, Car)\}$$

Notice the importance of knowing the source and target sets; we can conclude that Mike does not own any form of transport, while no-one out of Mike, Nasrin and Carlos owns a horse or a plane; the set of ordered pairs just on its own fails to capture all the information.

In principle, we have a special term, *correspondence*, to refer to the set of ordered pairs used to model a relation.

Definition 5.1 A correspondence between a source set *A* and a target set *B* is a subset of *A* × *B*.

In practice, because modelling a relation by a set of ordered pairs is so common, we usually use the term 'relation' to refer to the set itself. Thus we will usually speak of a relation *R* between *A* and *B* as being a subset of *A* × *B*, that is, an element of $\mathbb{P}(A \times B)$. This practice will be adopted in this book.

Notation 5.2 If an object *a* is related by *R* to another object *b* then we write *a R b*; since we regard a relation as a set of ordered pairs, then the notation *a R b* is regarded as an alternative to writing $(a, b) \in R$.

5.4.2   Higher order relations

An ordered pair is a particular example of what is often termed a *tuplet*. Thus we can have ordered triplets (of three elements), quadruplets (of four elements) and so on. A ternary relation would then be modelled by a set of triplets, a quaternary relation by a set of quadruplets and so on.

For example the relation ... *is_between* ... *and* ... might relate triples of railway stations along a line, for example:

Princeton *is_between* Trenton *and* Newark

This relation is an example of a ternary relation.

Although we sometimes do need higher order relations, binary relations are much more common. Because of this, we frequently refer to a binary relation simply as a 'relation'; this convention will be used in this book.

5.4.3   Diagrammatic models of relations

Diagrams can also be used to model relations, and are often used alongside the set models. Consider, for example, the relation R in which $1R2$, $2R1$, $2R3$, $2R4$ and $3R1$. This relation can be modelled by either a Cartesian plot (Figure 5.1) or an arrow diagram (Figure 5.2).

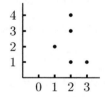

Figure 5.1 *An example of a Cartesian plot.*

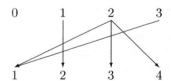

Figure 5.2 *An example of an arrow diagram.*

5.4.4   A simple example of a relation

Suppose we have two sets: $A = \{0, 1, 2\}$ and $B = \{1, 2, 3\}$. Then any relation, R, between A and B is a subset of $A \times B$. We indicate this by writing $R :\in \mathbb{P}(A \times B)$. From this declaration we can see that the source set for R is A and the target set is B.

To complete the definition of a particular relation, we need to give it a value. For example, one possible value for R is $\{(1, 1), (2, 1), (2, 2), (2, 3)\}$; another possible value is $\{(0, 1), (2, 1), (0, 2), (2, 3)\}$.

Examples 5.3

1. Write down two other possible values for $R :\in \mathbb{P}(A \times B)$.
2. How many different possible values for R are there in total?

Solution 5.4

1. Two further possible values include $\{\}$ and $\{(0, 3)\}$. Note that the *null* relation modelled by the empty set is as valid a relation as any other.
2. The product set $A \times B$ contains $3 * 3 = 9$ elements (each element being an ordered pair). The power set of $A \times B$ will therefore contain 2^9 elements with each element being a relation. Thus there are $2^9 = 512$ different possible values for R.

Notation 5.5 The notation $A \leftrightarrow B$ is frequently used as a shorthand for $\mathbb{P}(A \times B)$; it represents the set of all possible relations with source A and target B.

Thus in the example above $A \leftrightarrow B$ would contain 512 elements:

$$A \leftrightarrow B = \{\{\}, \{(0, 1)\}, \{(0, 1), (0, 2)\}, \ldots\}$$

Examples 5.6 Suppose that for the relation R, x is related to y if and only if all three of the following conditions are met:

1. x is a member of $\{1, 2, 3\}$;
2. y is a member of $\{1, 2, 3, 4\}$;
3. $x < y$.

What set can be used to model this relation?

Solution 5.7 From the given conditions we can see that the source set is $\{1, 2, 3\}$ and the target set is $\{1, 2, 3, 4\}$. The set modelling the relation is therefore a subset of $\{1, 2, 3\} \times \{1, 2, 3, 4\}$; it is equal to

$$\{ x :\in \{1, 2, 3\}; \ y :\in \{1, 2, 3, 4\} \mid x < y \bullet (x, y) \}$$

The set enumeration is $\{(1, 2), (1, 3), (1, 4), (2, 3), (2, 4), (3, 4)\}$.

5.5 Operations on relations considered as sets

In this, and subsequent sections, the term 'relation' will be used to refer to the set of ordered pairs corresponding to a binary relation.

5.5.1 Domain and range

The terms domain and range generalize easily to refer to relations.

Definition 5.8 The set of first coordinates is called the *domain* of the relation.

Notation 5.9 The domain of a relation R is denoted by $\operatorname{dom} R$.

Definition 5.10 The set of second coordinates is called the *range* of the relation.

Notation 5.11 The range of a relation R is denoted by $\operatorname{ran} R$.

Examples 5.12 Write down the domain and range for the following relations:

1. $R = \{(0, 1), (1, 3), (1, 6), (2, 4), (2, 5), (2, 0)\}$
2. $S = \{(0, 1), (1, 3), (2, 0), (2, 2), (2, 2)\}$
3. $\varnothing$

Note that we do not need to know the source or target sets to find the domain and range.

Solution 5.13

1. $\operatorname{dom} R = \{0, 1, 1, 2, 2, 2\} = \{0, 1, 2\}$
   $\operatorname{ran} R = \{1, 3, 6, 4, 5, 0\} = \{0, 1, 3, 4, 5, 6\}$
2. $\operatorname{dom} S = \{0, 1, 2, 2, 2\} = \{0, 1, 2\}$
   $\operatorname{ran} S = \{1, 3, 0, 2, 2\} = \{0, 1, 2, 3\}$
3. $\operatorname{dom} \varnothing = \operatorname{ran} \varnothing = \varnothing$

5.5.2 Set operations

Since a relation is a set of ordered pairs, we can apply all the normal set operations such as union $\cup$, intersection $\cap$ and power set $\mathbb{P}$ to relations. Sometimes the result of such an operation will be a relation; other times it is not.

Examples 5.14 Suppose there are two relations such that

$$R = \{(0, 1), (0, 2), (1, 1), (3, 5)\}$$
$$S = \{(2, 1), (3, 5)\}$$

Evaluate each of the following:

1. $R \cup S$
2. $R \cap S$
3. $R \setminus S$
4. $R \times S$
5. $\mathbb{P} S$

Again note that it is not necessary to know the source or target sets in order to carry out these operations.

Solution 5.15

1. $R \cup S = \{(0,1),(0,2),(1,1),(3,5),(2,1)\}$ which is also a relation.
2. $R \cap S = \{(3,5)\}$ which is also a relation.
3. $R \setminus S = \{(0,1),(0,2),(1,1)\}$ which is also a relation.
4. $R \times S = \{((0,1),(2,1)),((0,2),(2,1)),((1,1),(2,1)),((3,5),(2,1)),$
   $((0,1),(3,5)),((0,2),(3,5)),((1,1),(3,5)),((3,5),(3,5))\}$ which is also a relation.
5. $\mathbb{P}S = \{\{(2,1),(3,5)\},\{(2,1)\},\{(3,5)\},\{\}\}$ which cannot be a relation since the individual elements are sets rather than ordered pairs

5.5.3 Relational inverse

Definition 5.16 The inverse of a relation is obtained by swapping round the coordinates.

Notation 5.17 The inverse of a relation R is denoted by $R^\sim$.

Examples 5.18 Suppose R, S are two relations such that

$$
\begin{aligned}
R &= \{(0,1),(0,2),(1,1),(3,5)\} \\
S &= \{(2,1),(3,5)\}
\end{aligned}
$$

Evaluate each of the following:

1. $R^\sim$
2. $S^\sim$
3. $(R^\sim)^\sim$
4. $(S^\sim)^\sim$

Yet again we find that it is not necessary to know the source and target sets in order to apply relational inverse.

Solution 5.19

1. $R^\sim = \{(1,0),(2,0),(1,1),(5,3)\}$
2. $S^\sim = \{(1,2),(5,3)\}$
3. $(R^\sim)^\sim = \{(0,1),(0,2),(1,1),(3,5)\} = R$
4. $(S^\sim)^\sim = \{(2,1),(3,5)\} = S$

Note that

$$
\begin{aligned}
\operatorname{dom} R &= \operatorname{ran} R^\sim &= \{0,1,3\} \\
\operatorname{ran} R &= \operatorname{dom} R^\sim &= \{1,2,5\} \\
\operatorname{dom} S &= \operatorname{ran} S^\sim &= \{2,3\} \\
\operatorname{ran} S &= \operatorname{dom} S^\sim &= \{1,5\}
\end{aligned}
$$

Fact 5.20 In general, for any relation X it is true that $\operatorname{dom} X = \operatorname{ran} X^{\sim}$ and $\operatorname{ran} X = \operatorname{dom} X^{\sim}$.

Fact 5.21 In general, for any relation X it is true that $(X^{\sim})^{\sim} = X$.

5.5.4   Composition

Definition 5.22 The composition of R with S is equal to the set of all those ordered pairs (x, z) such that we can find at least one value of y such that xRy and ySz.

Notation 5.23 The relational composition of R with S is denoted by $R \, {}_9^\circ \, S$.

Notice that an alternative notation (the 'backward' notation) is sometimes used: instead of $R \, {}_9^\circ \, S$ we can write $S \circ R$ to mean the same thing. This will be considered further in Chapter 7.

Definition 5.24 More formally, if $R :\in A{\leftrightarrow}B$ and $S :\in B{\leftrightarrow}C$ then

$$R \, {}_9^\circ \, S \mathrel{\hat{=}} \{a :\in A; \; c :\in C \mid (\exists b :\in B \bullet aRb \wedge bRc) \bullet (a, c)\}$$

In practice, this definition tends to be a little cumbersome to use. An easier way is to use the arrow diagram.

Examples 5.25 Suppose $A = \{1, 2, 3\}$, $B = \{2, 3, 5, 7, 11\}$, $C = \{12, 16, 21\}$ and we define $R :\in A{\leftrightarrow}B$ and $S :\in B{\leftrightarrow}C$ by

$$R = \{(1, 2), (1, 3), (1, 11), (2, 3), (2, 5), (2, 7), (2, 11)\}$$

$$S = \{(2, 12), (2, 16), (3, 12), (3, 21), (7, 21)\}$$

What is the value of $R \, {}_9^\circ \, S$?

Solution 5.26 From the definition it would seem that we need to take each element x in A with each element z in C and find whether there is an element y in B such that $(x, y) \in R$ and $(y, z) \in S$. For example since $(1, 2) \in R$ and $(2, 12) \in S$ then $(1, 12) \in R \, {}_9^\circ \, S$. However, the answer can be more easily obtained by a diagram (see Figure 5.3).

Thus we see that

$$R \, {}_9^\circ \, S = \{(1, 12), (1, 16), (1, 21), (2, 12), (2, 21)\}$$

Note that it does not make sense to talk of $S \, {}_9^\circ \, R$ since the target set of S is not the source set of R.

We can, of course, combine the operations of inversion and composition provided the source and target sets are compatible.

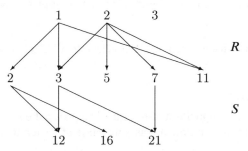

Figure 5.3 *Composition of relations.*

Examples 5.27 Suppose $R, S :\in \mathbb{N} \leftrightarrow \mathbb{N}$ are two relations such that

$$R = \{(0,0), (0,2), (1,1)\}$$
$$S = \{(1,2), (2,1)\}$$

Evaluate

1. $(R \,\mathbin{_9^9}\, S)^{\sim}$
2. $R^{\sim} \,\mathbin{_9^9}\, S^{\sim}$
3. $S^{\sim} \,\mathbin{_9^9}\, R^{\sim}$
4. $R \,\mathbin{_9^9}\, R^{\sim}$
5. $R^{\sim} \,\mathbin{_9^9}\, R$

Solution 5.28

1. $(R \,\mathbin{_9^9}\, S)^{\sim} = \{(0,1), (1,2)\} = \{(1,0), (2,1)\}$
2. $R^{\sim} \,\mathbin{_9^9}\, S^{\sim} = \{(0,0), (2,0), (1,1)\} \,\mathbin{_9^9}\, \{(2,1), (1,2)\} = \{(1,2)\}$
3. $S^{\sim} \,\mathbin{_9^9}\, R^{\sim} = \{(2,1), (1,2)\} \,\mathbin{_9^9}\, \{(0,0), (2,0), (1,1)\} = \{(2,1), (1,0)\}$
4. $R \,\mathbin{_9^9}\, R^{\sim} = \{(0,0), (0,2), (1,1)\} \,\mathbin{_9^9}\, \{(0,0), (2,0), (1,1)\} = \{(0,0), (1,1)\}$
5. $R^{\sim} \,\mathbin{_9^9}\, R = \{(0,0), (2,0), (1,1)\} \,\mathbin{_9^9}\, \{(0,0), (0,2), (1,1)\}$
   $= \{(0,0), (0,2), (2,0), (2,2), (1,1)\}$

Note carefully that in this last example, $R \,\mathbin{_9^9}\, R^{\sim} \neq R^{\sim} \,\mathbin{_9^9}\, R$ and that $(R \,\mathbin{_9^9}\, S)^{\sim} \neq R^{\sim} \,\mathbin{_9^9}\, S^{\sim}$, but that $(R \,\mathbin{_9^9}\, S)^{\sim} = S^{\sim} \,\mathbin{_9^9}\, R^{\sim}$.

Fact 5.29 No matter what values are chosen for R and S, $(R \,\mathbin{_9^9}\, S)^{\sim}$ and $S^{\sim} \,\mathbin{_9^9}\, R^{\sim}$ will always give the same result. We say that the two expressions are equivalent, and write

$$(R \,\mathbin{_9^9}\, S)^{\sim} \equiv S^{\sim} \,\mathbin{_9^9}\, R^{\sim}$$

5.5.5   Relational override

Definition 5.30 The override of R by S is denoted by $R \oplus S$ and is obtained by adding to S all those ordered pairs from R whose first coordinates are not in the domain of S.

Examples 5.31 If R, S are relations with values given by

$$R = \{(0, 1), (0, 2), (2, 3)\}$$
$$S = \{(0, 1), (1, 3), (3, 0)\}$$

evaluate:

1. $R \oplus S$
2. $S \oplus R$

Solution 5.32

1. $R \oplus S = \{(0, 1), (1, 3), (3, 0), (2, 3)\}$. The ordered pair $(2, 3)$ has been added since $2 \notin \operatorname{dom} S$, but the ordered pair $(0, 2)$ has not been added since $0 \in \operatorname{dom} S$; note that $(0, 1)$ is in both R and S.
2. $S \oplus R = \{(0, 1), (0, 2), (2, 3), (1, 3), (3, 0)\}$.

5.6   Some special types of relation

We can identify special kinds of relation, often with special properties.

5.6.1   Null relation

Definition 5.33 The *null relation* is equal to the empty set, $\varnothing$.

The null relation represents the idea that nothing in the source set is related to anything in the target set – this may well be important information! For example if we consider the relation *is_partner_of* between the set *Examiners* (for a certain examination) and the set *Candidates* (for that examination), then it might be reassuring to know that the relation is indeed empty.

5.6.2   Universal relation

At the other extreme to the null relation is the universal relation, when *everything* in the source set A is related to everything in the target set B.

Definition 5.34 The universal relation is equal to the product of the source and target sets, $A \times B$.

5.6.3   Identity relation

Definition 5.35 The *identity relation* on a set A is the set of ordered pairs of each element of A coupled with itself:

$$\{ x :\in A \bullet (x, x) \}$$

Notation 5.36 The identity relation on a set A will be denoted by I_A or $\mathrm{id}\, A$.

Examples 5.37 If $X = \{2, 4, 6\}$ and $Y = \{0, 1\}$ write down:

1. the universal relation from X to Y;
2. the universal relation from Y to X;
3. the identity relation, I_X, on X;
4. the identity relation, I_Y, on Y.

Solution 5.38

1. The universal relation from X to Y is given simply by $X \times Y$, that is

$$\{(2, 0), (2, 1), (4, 0), (4, 1), (6, 0), (6, 1)\}$$

2. The universal relation from Y to X will be different to that from X to Y. Its value is

$$\{(0, 2), (0, 4), (0, 6), (1, 2), (1, 4), (1, 6)\}$$

which is in fact $(X \times Y)^{\sim}$.
3. $I_X = \{(2, 2), (4, 4), (6, 6)\}$.
4. $I_Y = \{(0, 0), (1, 1)\}$.

5.6.4   One-to-one correspondences

Definition 5.39 If, in a correspondence, each element of the source set is uniquely associated with an element of the target set, while each element of the target set is uniquely associated with an element of the source set, then that relation is said to be a *one-to-one correspondence*.

Notation 5.40 We often write 1-1 correspondence.

The concept is illustrated in Figure 5.4.

Fact 5.41 If R is a 1-1 correspondence then so is $R^{\sim}$.

From the diagram it can also be seen that the composition of a 1-1 correspondence with its inverse will always give an identity relation; this gives us an alternative definition.

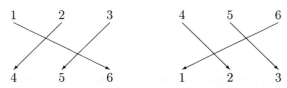

Figure 5.4 *An example of a 1-1 correspondence and its inverse.*

Definition 5.42 A relation R with source set A and target set B is a 1-1 correspondence if and only if both $R \,\mathring{,}\, R^{\sim} = I_A$ and $R^{\sim} \,\mathring{,}\, R = I_B$ are true.

Examples 5.43 Which of the following relations are 1-1 correspondences?

1. $\{(1,6),(2,5),(3,7)\}$ with source set $\{1,2,3\}$ and target set $\{5,6,7\}$.
2. $\{(1,6),(2,5),(3,5)\}$ with source set $\{1,2,3\}$ and target set $\{5,6,7\}$.
3. $\{(1,6),(2,5),(2,7)\}$ with source set $\{1,2,3\}$ and target set $\{5,6,7\}$.
4. $\{(1,6),(2,5),(3,7)\}$ with source set $\{1,2,3\}$ and target set $\{5,6,7,8\}$.

Solution 5.44

1. Checking *both* conditions:

   - $\{(1,6),(2,5),(3,7)\} \,\mathring{,}\, \{(1,6),(2,5),(3,7)\}^{\sim} = \{(1,1),(2,2),(3,3)\}$
     $= I_{\{1,2,3\}}$
   - $\{(1,6),(2,5),(3,7)\}^{\sim} \,\mathring{,}\, \{(1,6),(2,5),(3,7)\} = \{(6,6),(5,5),(7,7)\}$
     $= I_{\{5,6,7\}}$

   In this case both conditions are met and hence the given relation is indeed a 1-1 correspondence.

2. Checking *both* conditions:

   - $\{(1,6),(2,5),(3,5)\} \,\mathring{,}\, \{(1,6),(2,5),(3,5)\}^{\sim}$
     $= \{(1,1),(2,2),(3,3),(2,3),(3,2)\} \neq I_{\{1,2,3\}}$
   - $\{(1,6),(2,5),(3,5)\}^{\sim}\,\mathring{,}\,\{(1,6),(2,5),(3,5)\} = \{(6,6),(5,5)\} \neq I_{\{5,6,7\}}$

   We see that neither are true in this case so that the given relation is not a 1-1 correspondence.

3. Checking *both* conditions:

   - $\{(1,6),(2,5),(2,7)\}\,\mathring{,}\,\{(1,6),(2,5),(2,7)\}^{\sim} = \{(1,1),(2,2)\} \neq I_{\{1,2,3\}}$
   - $\{(1,6),(2,5),(2,7)\}^{\sim} \,\mathring{,}\, \{(1,6),(2,5),(2,7)\}$
     $= \{(6,6),(5,5),(7,7),(7,5),(5,7)\} \neq I_{\{5,6,7\}}$

   We see that neither are true in this case so that the given relation is not a 1-1 correspondence.

4. Checking *both* conditions:

- $\{(1,6),(2,5),(3,7)\} \,\S\, \{(1,6),(2,5),(3,7)\}^{\sim} = \{(1,1),(2,2),(3,3)\}$
  $= I_{\{1,2,3\}}$
- $\{(1,6),(2,5),(3,7)\}^{\sim} \,\S\, \{(1,6),(2,5),(3,7)\} = \{(6,6),(5,5),(7,7)\}$
  $\neq I_{\{5,6,7,8\}}$

We see that although the first condition is met in this case, the second is not. Hence the given relation is not a 1-1 correspondence. In fact there can be no 1-1 correspondence with the given source and target sets since their cardinalities are different.

Fact 5.45 The cardinalities of source and target sets in a 1-1 correspondence must be equal.

Note however that even when the sets do have equal cardinalities, it does not necessarily follow that a relation between them is a 1-1 correspondence.

5.7 Case study: project teams

We continue with the case study on project teams introduced in section 2.4. Recall that we had a set *Workforce* equal to $\{Elma, Rajesh, Mary, Carlos, Mike\}$ from which we took four project teams: *project_A*, *project_B*, *project_C* and *project_D*. In this section we are going to look at how relations can be used to represent information about this workforce and how relational operations can be used to extract further information.

When we collect *data*, perhaps from a survey questionnaire or from existing files, we get a collection of facts and figures. Organizing these facts and figures in some meaningful way results in *information*. One possible way of organizing data into information is to use relations; this is essentially the idea behind relational databases and indeed the language *PROLOG*.

For example look at the information presented in section 2.4. This information has been presented in the form of sets: apart from an overall set *Workforce*, four further sets are required to define the project teams. The project team information is thus stored as four separate mathematical entities. This way of storing the information is fine for noting changes in the project teams themselves: the changed information is reflected simply by altering the values of the sets involved. If, however, project teams are closed down or new ones started, it is necessary to add or remove the entities which comprise the information rather than simply alter their values. How much better it would be to hold all the information in a single entity whose value can be changed to represent *any* alteration in the deployment of staff to projects. This can be done by establishing a relation *project_member*. This relation will be (modelled by) a set of ordered pairs. There is, however, some degree of choice as to what to choose for the source and target sets.

We could take the source set to be *Workforce* and the target set to be the set of all four projects {*project_A, project_B, project_C, project_D*}. This has the disadvantage that the restriction of the target set to existing project teams prevents us from adding new teams (though teams can be removed). Indeed the same problem applies to using *Workforce* as the source set; it prevents us from adding new members of staff.

It would seem to be much better to introduce the target set *All_Project* which has all possible (past, current and future) project teams as its members. To simplify information storage we can now label project teams simply as letters, since the fact that the letters refer to project teams is implicit in the definition of the target set. Thus the set of current project teams would be *Projects* = {*A, B, C, D*}. Similarly the source set should be *People* to include everyone who might be employed.

We now have:

$$project\_member \quad :\in \quad People \leftrightarrow All\_Projects$$
$$project\_member = \{ \quad (Elma, B), (Elma, C),$$
$$(Rajesh, A), (Rajesh, B), (Rajesh, D),$$
$$(Mary, B), (Mary, D),$$
$$(Carlos, A), (Carlos, C), (Carlos, D),$$
$$(Mike, A), (Mike, B), (Mike, D)\}$$

We can now extract further information from this relation, for example:

- *Workforce* is given by the domain: *Workforce* = dom *project_member*.
- The set of current projects, *Projects*, is given by the range:

$$Projects = \text{ran } project\_member$$

- Suppose we wish to know the relation *common_project* between personnel who work together on at least one project team. This new relation can be obtained using the operation of relational composition:

$$common\_project = project\_member \, \S \, project\_member^{\sim}$$

For the *particular* example given, the resulting relation is equal to the product set *Workforce* × *Workforce*. Everyone works on at least one project with every other member of the workforce.

- Similarly we can extract information about projects which have at least one person in common:

$$common\_member = project\_member^{\sim} \, \S \, project\_member$$

in which we have used both relational composition and relational inversion. For the particular example given,

$$common\_member = \{(A, A), (A, B), (A, D),$$
$$(B, B), (B, D), (B, A), (B, C),$$
$$(C, C), (C, B), (C, A), (C, D),$$
$$(D, D), (D, A), (D, B), (D, C)\}$$

Inevitably, a project will have all its members in common with itself; perhaps a more useful definition of *common_member* would remove all terms of the form (x, x). This we shall be able to do once we have looked at *homogeneous relations* in Chapter 6.

Relational databases utilize a wide variety of relational operations to extract information.

Note that it is not necessary to store the values of *Workforce* and *Projects* separately since they can be extracted from the relation. Adopting this decision, however, leads to further difficulties. Suppose a new member of staff, *Helga*, is employed. Initially she may not be allocated to any project team, and so will not be included in the relation as it now stands; thus *Helga* would not be a member of dom *Projects*. One possible solution to this problem is to have a 'null' project, Z say, to which anyone is automatically allocated if they are not currently allocated to a real project. The set of current (real) projects would then be given by:

$$Projects = (\text{ran } project\_member) \setminus \{Z\}$$

We shall return to this point in section 7.7 where another possible solution is considered.

5.8   Exercise

In the following questions take $X = \{0, 1, 2, 3\}$, $Y = \{4, 5, 6\}$ and $Z = \{7, 8\}$. In addition, suppose:

- $R :\in X \leftrightarrow Y$ is equal to $\{(0, 4), (0, 5), (1, 5), (1, 6), (3, 4)\}$,
- $S :\in X \leftrightarrow Z$ is equal to $\{(0, 7), (1, 7), (2, 7), (3, 7), (2, 8)\}$,
- $T :\in Y \leftrightarrow Z$ is equal to $\{(4, 7), (4, 8), (5, 7)\}$,
- $U :\in Y \leftrightarrow X$ is equal to $\{(4, 2), (5, 2), (6, 2)\}$ and
- $V :\in Z \leftrightarrow Z$ is equal to $\{(7, 8)\}$.

As originally defined in section 2.4 take

$$Workforce = \{Elma, Rajesh, Mary, Carlos, Mike\}$$

with allocation of personnel to projects as shown in that section. Suppose also that this *Workforce* is to be allocated to offices *Big* and *Little* as follows:

- *Elma*, *Rajesh* and *Mary* share *Big*;
- *Mike* and *Carlos* share *Little*.

1. Evaluate

(a) I_X	(g) ran S	(m) ran V
(b) I_Y	(h) dom T	(n) $R^\sim$
(c) I_Z	(i) ran T	(o) $S^\sim$
(d) dom R	(j) dom U	(p) $T^\sim$
(e) ran R	(k) ran U	(q) $U^\sim$
(f) dom S	(l) dom V	(r) $V^\sim$

2. Evaluate

(a) $R \mathbin{\substack{\circ\\\circ}} I_Y$	(d) $T \mathbin{\substack{\circ\\\circ}} V$	(g) $(R \mathbin{\substack{\circ\\\circ}} T) \mathbin{\substack{\circ\\\circ}} V$
(b) $I_X \mathbin{\substack{\circ\\\circ}} R$	(e) $S \mathbin{\substack{\circ\\\circ}} V$	(h) $R \mathbin{\substack{\circ\\\circ}} (T \mathbin{\substack{\circ\\\circ}} V)$
(c) $R \mathbin{\substack{\circ\\\circ}} T$	(f) $R \mathbin{\substack{\circ\\\circ}} U$	(i) $V \mathbin{\substack{\circ\\\circ}} V$

3. Evaluate

(a) $R \mathbin{\substack{\circ\\\circ}} R^\sim$	(e) $T \mathbin{\substack{\circ\\\circ}} T^\sim$	(i) $V \mathbin{\substack{\circ\\\circ}} V^\sim$
(b) $R^\sim \mathbin{\substack{\circ\\\circ}} R$	(f) $T^\sim \mathbin{\substack{\circ\\\circ}} T$	(j) $V^\sim \mathbin{\substack{\circ\\\circ}} V$
(c) $S \mathbin{\substack{\circ\\\circ}} S^\sim$	(g) $U \mathbin{\substack{\circ\\\circ}} U^\sim$	(k) $(R \mathbin{\substack{\circ\\\circ}} T) \mathbin{\substack{\circ\\\circ}} S^\sim$
(d) $S^\sim \mathbin{\substack{\circ\\\circ}} S$	(h) $U^\sim \mathbin{\substack{\circ\\\circ}} U$	(l) $R \mathbin{\substack{\circ\\\circ}} (T \mathbin{\substack{\circ\\\circ}} S^\sim)$

4. Define a relation, *accommodates*, between offices and people to capture information on office allocation; take *Offices* as the source set and *People* as the target set. Evaluate each of the following expressions and explain what each expression represents:

   (a) *accommodates*

   (b) dom *accommodates*

   (c) ran *accommodates*

   (d) *accommodates*$^\sim$

   (e) *accommodates* $\mathbin{\substack{\circ\\\circ}}$ *accommodates*

   (f) *accommodates*$^\sim$ $\mathbin{\substack{\circ\\\circ}}$ *accommodates*$^\sim$

   (g) *accommodates* $\mathbin{\substack{\circ\\\circ}}$ *project_member*

5. Instead of having two separate *binary* relations to hold information on project and office allocation, it would be possible to have a single *ternary* relation between *Offices*, *People* and *Projects*:

   $$Allocations :\in \mathbb{P}(Offices \times People \times Projects)$$

   What is the value of *Allocations*? What are the disadvantages of using this one ternary relation instead of the two separate binary relations?

CHAPTER 6

Homogeneous relations

6.1   What is a homogeneous relation?

In a binary relation, the source and target sets will in general be different. There are many situations however when the source and target sets are the *same*.

Definition 6.1 A homogeneous binary relation is one in which the source and target sets are identical.

Fact 6.2 A homogeneous relation from a set X to X is often spoken of as a *relation on X*.

Examples 6.3 Suppose we have two different sets, *Names* and *Places*, and further suppose relations R, S and T are defined with
$R :\in$ *Names↔Names*
$S :\in$ *Names↔Places* and
$T :\in$ *Places↔Names*.
Which of the following are homogeneous relations, and on what set are they relations?

   1. R;
   2. S;
   3. T;
   4. $R^\sim$;
   5. $S \mathbin{\stackrel{\circ}{\scriptscriptstyle 9}} S^\sim$;
   6. $R \mathbin{\stackrel{\circ}{\scriptscriptstyle 9}} S$;
   7. $S \mathbin{\stackrel{\circ}{\scriptscriptstyle 9}} T$;
   8. $T \mathbin{\stackrel{\circ}{\scriptscriptstyle 9}} R \mathbin{\stackrel{\circ}{\scriptscriptstyle 9}} S$.

Solution 6.4

   1. R has *Names* for both source and target sets, and is hence a homogeneous relation on *Names*;
   2. S is a relation from *Names* to *Places* and is hence not homogeneous;
   3. T is a relation from *Places* to *Names* and is hence not homogeneous;

4. $R^\sim$ is homogeneous – the inverse of any homogeneous relation will always be homogeneous on the same set;
5. $S \, \mathring{,} \, S^\sim$ is a new relation with source set and target set both equal to *Names*, and is hence homogeneous on *Names* – the composition of any relation with its inverse will always be homogeneous;
6. $R \, \mathring{,} \, S$ is a relation from *Names* to *Places*;
7. $S \, \mathring{,} \, T$ is homogeneous on *Names*;
8. $T \, \mathring{,} \, R \, \mathring{,} \, S$ is homogeneous on *Places*.

Homogeneous relations are useful in many different types of circumstance. Two examples are

1. A family tree is a diagrammatic representation for several relations on a set of people, such as *parent_of* and *married_to*.
2. If several bus companies operate in the same area, then we may be interested in the relation *same_company* to relate places which are served by a common bus company.

6.1.1 Identity relations

We have already met one category of homogeneous relation – the identity relation on a set. Recall that for a given set X the *identity relation*, often called the *identity function*, is such that every element of X is related to itself and only itself.

Definition 6.5 For a set X the identity relation I_X is given by

$$I_X = \{x :\in X \bullet (x,x)\}$$

Examples 6.6 What is the identity relation on the set $\{0, 1, 2\}$?

Solution 6.7 The identity relation on $\{0, 1, 2\}$ is $\{(0,0), (1,1), (2,2)\}$.

6.1.2 Diagrammatic representation

The diagrams used for any relation may of course be used for homogeneous relation. However, a different layout may sometimes be used. The resulting form of the diagram is often called a *directed graph* or just *digraph*. (Strictly speaking it is a diagrammatic representation of a digraph.)

Examples 6.8 Suppose we have a relation R with source and target sets both equal to $\{1, 2, 3, 4, 5\}$ and that R is equal to

$$\{(1,2), (2,1), (2,2), (2,4), (2,5), (3,1), (3,5), (4,5), (5,5)\}$$

Represent this relation by a digraph.

Solution 6.9 Figure 6.1 is a digraph representation of the relation.

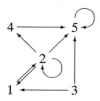

Figure 6.1 *Digraph for* $\{(1,2),(2,1),(2,2),(2,4),(2,5),(3,1),(3,5),(4,5),(5,5)\}$.

6.2   Classes of homogeneous relations

Homogeneous relations often have special properties which, apart from being useful information in their own right, might enable us to store data more efficiently.

In general it is *not* true that $R^\smallsmile = R$. This property will be true only for certain relations. In fact it could only ever be true for homogeneous relations since in order for a relation to equal its inverse, the source and target sets must be the same. But it is not even the case that all homogeneous relations will share this property.

The property $R^\smallsmile = R$ in fact categorizes a special class of homogeneous relations. Similarly other properties categorize other classes of homogeneous relations; we shall now look at some of these.

6.2.1   Reflexive relations

If a relation is *reflexive* then *every* element of the set is related to itself.

Definition 6.10 A relation R on a set X is reflexive if and only if it has the following property:

$$\forall x :\in X \bullet (x,x) \in R$$

Alternatively:

Definition 6.11 A relation R on a set X is reflexive if and only if

$$I_X \subseteq R$$

where I_X is the identity relation.

Examples 6.12 Which of the following relations on $\{1,2,3\}$ are reflexive?

1. $\{(1,1),(1,2),(2,2),(3,3),(3,1)\}$
2. $\{(1,1),(2,2)\}$

Solution 6.13   The identity relation I on $\{1,2,3\}$ is $\{(1,1),(2,2),(3,3)\}$.

1. $I \subseteq \{(1,1),(1,2),(2,2),(3,3),(3,1)\}$ which is therefore reflexive.
2. $(3,3) \notin \{(1,1),(2,2)\}$ which is therefore not reflexive.

6.2.2 Symmetric relations

If whenever b is related to a then a is related to b, the relation is said to be *symmetric*.

Definition 6.14 A relation R on X is symmetric if and only if

$$\forall x, y :\in X \bullet (x, y) \in R \Leftrightarrow (y, x) \in R$$

or, equivalently,

Definition 6.15 A relation R on X is symmetric if and only if $R^{\sim} = R$.

Examples 6.16 Which of the following relations on $\{1, 2, 3\}$ are symmetric?

1. $\{(1, 1), (1, 2), (2, 1), (2, 3), (3, 2), (3, 3)\}$
2. $\{(1, 1), (2, 2)\}$
3. $\{(1, 1), (2, 2), (3, 3), (3, 1)\}$

Solution 6.17

1. The inverse relation, $\{(1, 1), (1, 2), (2, 1), (2, 3), (3, 2), (3, 3)\}^{\sim}$, is equal to the original relation $\{(1, 1), (1, 2), (2, 1), (2, 3), (3, 2), (3, 3)\}$ which is therefore symmetric.
2. $\{(1, 1), (2, 2)\}^{\sim} = \{(1, 1), (2, 2)\}$ which is therefore symmetric.
3. $\{(1, 1), (2, 2), (3, 3), (3, 1)\}^{\sim} = \{(1, 1), (2, 2), (3, 3), (1, 3)\}$. This is not equal to $\{(1, 1), (2, 2), (3, 3), (3, 1)\}$ and hence the given relation is not symmetric.

6.2.3 Transitive relations

If whenever a is related to b and b is related to c then a is related to c, the relation is said to be *transitive*.

Definition 6.18 A relation R on X is transitive if and only if

$$\forall x, y, z :\in X \bullet (x, y) \in R \wedge (y, z) \in R \Rightarrow (x, z) \in R$$

Alternatively,

Definition 6.19 A relation R on X is transitive if and only if

$$R \, {}_9^9 R \subseteq R$$

Examples 6.20 Which of the following relations on $\{1, 2, 3\}$ are transitive?

1. $\{(1, 1), (1, 2), (2, 3), (1, 3), (3, 3)\}$
2. $\{(1, 1), (2, 3)\}$
3. $\{(1, 2), (2, 3), (3, 1)\}$

Solution 6.21

1. $\{(1,1),(1,2),(2,3),(1,3),(3,3)\} \, \raise0.3ex\hbox{$\scriptstyle\circ$}\kern-0.2em\raise-0.3ex\hbox{$\scriptstyle\circ$} \, \{(1,1),(1,2),(2,3),(1,3),(3,3)\}$
   $= \{(1,1),(1,2),(2,3),(1,3),(3,3)\}$
   $\subseteq \{(1,1),(1,2),(2,3),(1,3),(3,3)\}$.
   Hence we conclude that the given relation is transitive.
2. $\{(1,1),(2,3)\} \, \raise0.3ex\hbox{$\scriptstyle\circ$}\kern-0.2em\raise-0.3ex\hbox{$\scriptstyle\circ$} \, \{(1,1),(2,3)\} = \{(1,1)\} \subseteq \{(1,1),(2,3)\}$.
   The given relation is therefore transitive.
3. $\{(1,2),(2,3),(3,1)\} \, \raise0.3ex\hbox{$\scriptstyle\circ$}\kern-0.2em\raise-0.3ex\hbox{$\scriptstyle\circ$} \, \{(1,2),(2,3),(3,1)\}$
   $= \{(1,3),(2,1),(3,2)\}$
   $\not\subseteq \{(1,2),(2,3),(3,1)\}$.
   The given relation is therefore not transitive.

6.2.4   Equivalence relations

Definition 6.22 If a relation is reflexive, symmetric and transitive, then it is called an *equivalence* relation.

Examples 6.23 Which of the following relations on $\{1,2,3\}$ are equivalence relations?

1. $\{(1,1),(1,2),(2,1),(2,2),(3,3)\}$
2. $\{(1,1),(2,2),(3,3)\}$
3. $\{(1,2),(2,3),(3,1)\}$

Solution 6.24

1. $\{(1,1),(1,2),(2,1),(2,2),(3,3)\}$ is reflexive, symmetric and transitive; it is therefore an equivalence relation.
2. $\{(1,1),(2,2),(3,3)\}$ is reflexive, symmetric and transitive; it is therefore an equivalence relation.
3. $\{(1,2),(2,3),(3,1)\}$ is transitive but is neither reflexive nor symmetric; it is therefore not an equivalence relation.

Notice how the equivalence relation $\{(1,1),(1,2),(2,1),(2,2),(3,3)\}$ can be written as $(\{1,2\} \times \{1,2\}) \cup (\{3\} \times \{3\})$. Furthermore the source set $\{1,2,3\}$ can be written as $\{1,2\} \cup \{3\}$.

We say that $\{\{1,2\},\{3\}\}$ constitutes a *partition* of $\{1,2,3\}$ and that this partition is associated with the equivalence relation $(\{1,2\} \times \{1,2\}) \cup (\{3\} \times \{3\})$.

Examples 6.25 Which of the following are partitions of $\{1,2,3\}$?

1. $\{1,2,3\}$
2. $\{\{1\},\{2\},\{3\}\}$
3. $\{\{1,2,3\}\}$
4. $\{\{1,3\},\{2\}\}$

5. $\{\{1,3\}\}$
6. $\{\{1,3\},\{2,3\}\}$
7. $\{\{1,3\},\{2\},\varnothing\}$

Solution 6.26

1. A partition of $\{1,2,3\}$ is a subset of the power set $\mathbb{P}\{1,2,3\}$. Now $\{1,2,3\}$ is an element of $\mathbb{P}\{1,2,3\}$, not a subset, and so is not a partition.
2. Each element of $\{1,2,3\}$ occurs in precisely one element of $\{\{1\},\{2\},\{3\}\}$ which is therefore a partition.
3. Likewise $\{\{1,2,3\}\}$ is also a partition according to the definition.
4. $\{\{1,3\},\{2\}\}$ is also a partition.
5. $\{\{1,3\}\}$ is not a partition since 2 is absent from all the elements of the partition.
6. $\{\{1,3\},\{2,3\}\}$ is not a partition since 3 occurs twice: $\{1,3\}\cap\{2,3\}=\{3\}$.
7. $\{\{1,3\},\{2\},\varnothing\}$ is not a partition as one of the elements is the empty set.

Fact 6.27 Given a partition of a set X we can associate an equivalence relation; this equivalence relation is equal to the union of the products of each subset of X with itself.

Examples 6.28 What equivalence relation corresponds to each of the following partitions?

1. $\{\{1\},\{2\},\{3\}\}$
2. $\{\{1,2,3\}\}$
3. $\{\{1,3\},\{2\}\}$

Solution 6.29

1. $\bigcup\{\{1\}\times\{1\},\{2\}\times\{2\},\{3\}\times\{3\}\}$
   $=\bigcup\{\{(1,1)\},\{(2,2)\},\{(3,3)\}\}$
   $=\{(1,1),(2,2),(3,3)\}$
2. $\bigcup\{\{1,2,3\}\times\{1,2,3\}\}$
   $=\{1,2,3\}\times\{1,2,3\}$
   $=\{(1,1),(1,2),(1,3),(2,1),(2,2),(2,3),(3,1),(3,2),(3,3)\}$
3. $\bigcup\{\{1,3\}\times\{1,3\},\{2\}\times\{2\}\}=\{(1,1),(1,3),(3,1),(3,3),(2,2)\}$

Does the converse hold? Can we find a partition for every equivalence relation?

Fact 6.30 With every equivalence relation there is an associated partition.

Definition 6.31 The members of the partition associated with an equivalence relation are called *equivalence classes*.

Examples 6.32 What are the equivalence classes associated with each of the following equivalence relations on $\{0,2,4\}$?

1. $\{(0,0),(2,2),(4,4)\}$
2. $\{(0,0),(2,2),(4,4),(0,2),(2,0)\}$
3. $\{(0,0),(2,2),(4,4),(0,2),(2,0),(0,4),(4,0),(2,4),(4,2)\}$

Solution 6.33

1. No element is related to any element other than itself, and so each equivalence class contains just one element: $\{0\}$, $\{2\}$, $\{4\}$.
2. 4 is related to only itself, but 0 and 2 are related to each other as well as themselves. The equivalence classes are $\{0,2\}$ and $\{4\}$.
3. Each element is related to every other element; there is only one equivalence class, namely $\{0,2,4\}$.

6.2.5   Some other kinds of relation

Antisymmetric relations

If *aRb* and *bRa* cannot both be true for distinct elements *a* and *b*, then the relation is said to be *antisymmetric*. For example, if R is a relation on $\{1,2,3\}$ then

$$\{(1,1),(1,2),(2,3),(3,3)\}$$

is antisymmetric, and so is

$$\{(1,1),(2,2)\}$$

but

$$\{(1,1),(2,2),(2,3),(3,2),(3,3)\}$$

is not.

More formally a relation R on X is antisymmetric if and only if

$$\forall x,y :\in X \bullet (x,y) \in R \land (y,x) \in R \Rightarrow x = y$$

or, equivalently,

$$R^{\sim} \cap R \subseteq I_X$$

6.3   Closures

If a relation is known to have a special property such as reflexivity or symmetry for example, then there is a certain amount of redundancy in the information carried by that relation.

Examples 6.34 Find a subset of each of the following relations on $\{1,2\}$ which conveys the same information, given the special property:

1. R is reflexive with $R = \{(1,1),(2,2),(1,2)\}$
2. S is symmetric with $S = \{(1,2),(2,1),(2,2)\}$

Solution 6.35

1. We can remove the identity relation from R without loss of information:

$$R_{reduced} = R \setminus I_R = \{(1,2)\}$$

2. Only one of $(1,2)$ and $(2,1)$ is needed since the other can always be deduced:

$$S_{reduced} = \{(1,2),(2,2)\}$$

Given a reduced relation, the process of calculating the full relation is known as *closure*. Applying a closure operation to a homogeneous relation will always ensure the resulting new relation will have the corresponding properties.

Definition 6.36 For any homogeneous relation R on a set X, the *reflexive closure* is given by $R \cup I_X$.

Definition 6.37 For any homogeneous relation R on a set X, the *symmetric closure* is given by $R \cup R^{\sim}$.

Examples 6.38 Find the reflexive, symmetric, and symmetric–reflexive closures of the following relations on the set $X = \{1,2,3\}$:

1. $R = \{(1,1),(1,3),(2,1)\}$
2. $S = \{(1,2),(1,3),(2,1)\}$
3. $T = \{(1,1)\}$
4. $\varnothing$

Solution 6.39 To find the reflexive closures, we need to evaluate I_X:

$$I_X = \{(1,1),(2,2),(3,3)\}$$

1. The reflexive closure is

$$R \cup I_X = \{(1,1),(1,3),(2,1),(2,2),(3,3)\}$$

while the symmetric closure is

$$R \cup R^{\sim} = \{(1,1),(1,3),(2,1),(3,1),(1,2)\}$$

The symmetric–reflexive closure is the union of these two closures:

$$R \cup I_X \cup R^{\sim} = \{(1,1),(1,3),(3,1),(2,1),(1,2),(2,2),(3,3)\}$$

2. The reflexive closure is

$$\{(1,2),(1,3),(2,1),(1,1),(2,2),(3,3)\}$$

while the symmetric closure is

$$\{(1,2),(1,3),(2,1),(3,1)\}$$

and the symmetric–reflexive closure is

$$\{(1,2),(1,3),(2,1),(3,1),(1,1),(2,2),(3,3)\}$$

3. T is already symmetric and is hence equal to its symmetric closure. The reflexive and symmetric–reflexive closures are both equal to

$$\{(1,1),(2,2),(3,3)\}$$

4. $\varnothing$ is already symmetric and is hence equal to its symmetric closure. The reflexive and symmetric–reflexive closures are both equal to

$$\{(1,1),(2,2),(3,3)\}$$

One particularly important closure is transitive closure, and will be considered in a little more detail.

6.3.1   Transitive closure

Before explaining transitive closure, it is necessary to introduce a useful notation for repeated composition.

Examples 6.40 If $R = \{(1,2),(2,3),(3,2)\}$ calculate

1. $R \mathbin{\substack{\circ \\ 9}} R$
2. $R \mathbin{\substack{\circ \\ 9}} (R \mathbin{\substack{\circ \\ 9}} R)$
3. $(R \mathbin{\substack{\circ \\ 9}} R) \mathbin{\substack{\circ \\ 9}} R$

Solution 6.41

1. $R \mathbin{\substack{\circ \\ 9}} R = \{(1,3),(2,2),(3,3)\}$
2. $R \mathbin{\substack{\circ \\ 9}} (R \mathbin{\substack{\circ \\ 9}} R)$
   $= \{(1,2),(2,3),(3,2)\} \mathbin{\substack{\circ \\ 9}} \{(1,3),(2,2),(3,3)\}$
   $= \{(1,2),(2,3),(3,2)\}$
3. $(R \mathbin{\substack{\circ \\ 9}} R) \mathbin{\substack{\circ \\ 9}} R$
   $= \{(1,3),(2,2),(3,3)\} \mathbin{\substack{\circ \\ 9}} \{(1,2),(2,3),(3,2)\}$
   $= \{(1,2),(2,3),(3,2)\}$

Notice that the last two results are equal.

Fact 6.42 Relational composition of relations is associative, that is for *any* homogeneous relation:

$$R \mathbin{\raise0.3ex\hbox{$\scriptstyle\circ$}\kern-0.3em\raise-0.3ex\hbox{$\scriptstyle\circ$}} (R \mathbin{\raise0.3ex\hbox{$\scriptstyle\circ$}\kern-0.3em\raise-0.3ex\hbox{$\scriptstyle\circ$}} R) \equiv (R \mathbin{\raise0.3ex\hbox{$\scriptstyle\circ$}\kern-0.3em\raise-0.3ex\hbox{$\scriptstyle\circ$}} R) \mathbin{\raise0.3ex\hbox{$\scriptstyle\circ$}\kern-0.3em\raise-0.3ex\hbox{$\scriptstyle\circ$}} R$$

which we can therefore write unambiguously as

$$R \mathbin{\raise0.3ex\hbox{$\scriptstyle\circ$}\kern-0.3em\raise-0.3ex\hbox{$\scriptstyle\circ$}} R \mathbin{\raise0.3ex\hbox{$\scriptstyle\circ$}\kern-0.3em\raise-0.3ex\hbox{$\scriptstyle\circ$}} R$$

Notation 6.43 The composition of a relation with itself, $R \mathbin{\raise0.3ex\hbox{$\scriptstyle\circ$}\kern-0.3em\raise-0.3ex\hbox{$\scriptstyle\circ$}} R$, will be denoted by R^2. Similarly the composition of R with itself twice, that is $R \mathbin{\raise0.3ex\hbox{$\scriptstyle\circ$}\kern-0.3em\raise-0.3ex\hbox{$\scriptstyle\circ$}} R^2$ or $R \mathbin{\raise0.3ex\hbox{$\scriptstyle\circ$}\kern-0.3em\raise-0.3ex\hbox{$\scriptstyle\circ$}} R \mathbin{\raise0.3ex\hbox{$\scriptstyle\circ$}\kern-0.3em\raise-0.3ex\hbox{$\scriptstyle\circ$}} R$, will be denoted by R^3. In general, for $n > 1$, R^n will be used to denote $R \mathbin{\raise0.3ex\hbox{$\scriptstyle\circ$}\kern-0.3em\raise-0.3ex\hbox{$\scriptstyle\circ$}} R^{n-1}$; that is, R^n is the composition of R with itself $n - 1$ times.

Thus R^3 is equivalent to $R \mathbin{\raise0.3ex\hbox{$\scriptstyle\circ$}\kern-0.3em\raise-0.3ex\hbox{$\scriptstyle\circ$}} R^2 = R \mathbin{\raise0.3ex\hbox{$\scriptstyle\circ$}\kern-0.3em\raise-0.3ex\hbox{$\scriptstyle\circ$}} R \mathbin{\raise0.3ex\hbox{$\scriptstyle\circ$}\kern-0.3em\raise-0.3ex\hbox{$\scriptstyle\circ$}} R$. Likewise R^4 is equivalent to $R \mathbin{\raise0.3ex\hbox{$\scriptstyle\circ$}\kern-0.3em\raise-0.3ex\hbox{$\scriptstyle\circ$}} R^3 = R \mathbin{\raise0.3ex\hbox{$\scriptstyle\circ$}\kern-0.3em\raise-0.3ex\hbox{$\scriptstyle\circ$}} R \mathbin{\raise0.3ex\hbox{$\scriptstyle\circ$}\kern-0.3em\raise-0.3ex\hbox{$\scriptstyle\circ$}} R \mathbin{\raise0.3ex\hbox{$\scriptstyle\circ$}\kern-0.3em\raise-0.3ex\hbox{$\scriptstyle\circ$}} R$ and so on.

What is the meaning of repeated composition? The easiest way to interpret repeated composition is to think in terms of the digraph picture for the relation. For example suppose that we have the relation R as defined above. The digraph for R is shown in Figure 6.2.

Figure 6.2 *Digraph for $R = \{(1,2),(2,3),(3,2)\}$.*

The relation R itself corresponds to pairs of points the second of which can be reached in one 'jump' from the first; for example we can move from 1 to 2 along the *arc x*, or from 2 to 3 along arc *y*.

Now the relation R^2 is the set of pairs of values which can be linked by exactly two jumps. Thus we can move from 1 to 3 by arcs *x* and *y*; we can move from 2 along arc *y* then back to 2 again along arc *z*. Hence both $(1,3)$ and $(2,2)$ are elements of R^2. However, 1 and 2 cannot be linked in exactly two jumps, though they can in three jumps (and indeed in any odd number of jumps).

Similarly R^3 is the set of pairs linked by exactly three jumps, R^4 by exactly four jumps, and so on. Note that an ordered pair of values may be linked in more than one way; for example $(1, 2)$ is a member of $R, R^3, R^5, \ldots$.

For any general relation R we can define the concept of 'reachability'. If a point b is 'reachable' from a point a (perhaps the same point), then there must be some relation R^k for which $(a, b) \in R^k$; in other words b can be reached from a in k jumps. The relation to express reachability is the *transitive closure* of R.

Notation 6.44 The transitive closure of R is denoted by R^+.

If R is finite, then we can calculate R^+ as the union of terms of the form R^k:

$$R^+ = R \cup R^2 \cup R^3 \cup R^4 \cup \ldots \cup R^k \cup \ldots \cup R^n$$

where n is the maximum number of jumps necessary between reachable points.

For small relations, however, the simplest method of finding the transitive closure is by inspection of its digraph.

Examples 6.45 Find the transitive closure of

$$R = \{(0,\ 1)\ ,(1,\ 3)\ ,(3,\ 4)\ ,(4,\ 2)\ ,(4,\ 0)\}$$

Solution 6.46 The digraph for R is shown in Figure 6.3. From this it can be seen

Figure 6.3 *Digraph for $R = \{(0, 1)\ ,(1, 3)\ ,(3, 4)\ ,(4, 2)\ ,(4, 0)\}$.*

that we can go from 0 to 1 to 3 to 4; from 4 we can either go to 2 (which is a 'dead end') or back to 0. Hence from 0 we can reach $1, 3, 4, 2, 0$. Likewise from 1 we can reach $3, 4, 2, 0, 1$; from 3 we can reach $4, 2, 0, 1, 3$; from 4 we can reach $2, 0, 1, 3, 4$; it is not possible to move on from 2. Thus

$$
\begin{aligned}
R^+ \ = \ & \{(0,1)\ ,(0,3)\ ,(0,4)\ ,(0,2)\ ,(0,0)\ , \\
& (1,3)\ ,(1,4)\ ,(1,2)\ ,(1,0)\ ,(1,1)\ , \\
& (3,4)\ ,(3,2)\ ,(3,0)\ ,(3,1)\ ,(3,3)\ , \\
& (4,2)\ ,(4,0)\ ,(4,1)\ ,(4,3)\ ,(4,4)\}
\end{aligned}
$$

Transitive closure finds application in any area which can be represented by a digraph, such as networks and binary trees. In many cases, however, we also want each point to be related to itself. In the above example it could be argued that 2 is reachable from itself (in 0 jumps). This leads us to the idea of transitive–reflexive closure.

Transitive–reflexive closure

Notation 6.47 The transitive–reflexive closure of a relation R is denoted by R^*.

Definition 6.48 The transitive–reflexive closure of a relation R on a set X is given by

$$R^* = R^+ \cup I_X$$

Examples 6.49 If $X = \{0, 1, 2, 3, 4\}$ and $R = \{(0, 1)\ ,(1, 3)\ ,(3, 4)\ ,(4, 2)\ ,(4, 0)\}$, find R^*.

Solution 6.50 From the previous example we know the value of R^+. To obtain the corresponding reflexive relation, we must also have the element $(2,2)$ to give $R^*=$

$$\{(2,2),(0,1),(0,3),(0,4),(0,2),(0,0),(1,3),(1,4),(1,2),(1,0),(1,1),$$
$$(3,4),(3,2),(3,0),(3,1),(3,3),(4,2),(4,0),(4,1),(4,3),(4,4)\}$$

Note that if we regard the identity relation as relating points reachable in zero jumps, that is

$$R^0 = I_X$$

and the relation R itself as relating points reachable in one jump, that is

$$R^1 = R$$

then for finite relations we have

$$R^* = R^0 \cup R^1 \cup R^2 \cup R^3 \cup R^4 \cup \ldots \cup R^k \cup \ldots \cup R^n$$

for an appropriate value of n.

6.4   Case study: project teams

We have already met some examples of homogeneous relations in section 5.7, namely:

- *common_member = project_member~ ⨾ project_member =*
  $\{(A,A),(A,B),(A,D),(B,B),(B,D),(B,A),(B,C),$
  $(C,C),(C,B),(C,A),(C,D),(D,D),(D,A),(D,B),(D,C)\}$
  which is a relation on *All_Projects*.
- *common_project = project_member ⨾ project_member~ =*
  Workforce × Workforce which is a relation on *People*.

We notice some special properties of these relations.

- The relation *common_member* includes all the elements of the identity relation on its domain (in this case *Workforce*); it is also a symmetric relation.
- The relation *common_project* is in fact the universal relation on *Workforce*. Again, it includes all the elements of the identity relation on its domain (in this case *Projects*); it is also a symmetric relation.

This raises the following question:

As we alter the values of *common_member* and *common_project*, shall we always get symmetric relations which include all the elements of the identity relations on their domains?

The answer is yes! Both relations are of the form $R \,\S\, R^\sim$: for *common_project* take R as *project_member*; for *common_member* take R as *project_member*$^\sim$. Note that we have made use of a very important property of relations, namely:

Fact 6.51 $(R^\sim)^\sim = R$

Now it is also possible to prove the following very useful fact:

Fact 6.52 For *any* relation R whatsoever, both the relations

- $R \,\S\, R^\sim$
- $R^\sim \,\S\, R$

will always be symmetric.

Thus if we know this property to be always true, it will certainly be true for the special cases of *common_project* and *common_member*.

The ability to use general results such as Facts 6.51 and 6.52 in this manner is a very important aspect of the application of discrete mathematics. It requires us to be able to *reason* logically and to *prove* mathematical theorems. It is a very large topic in its own right, and requires more room than this book would allow; an excellent introduction is given in *A Logical Approach to Discrete Math* written by Gries and Schneider and published by Springer–Verlag.

6.5 Exercise

1. Decide whether each of the following relations on $\{0, 1, 2\}$ is reflexive, symmetric, antisymmetric, transitive, or a combination of these. Which are equivalence relations, and what are the corresponding equivalence classes?

    (a) $\{(0,0), (1,1), (1,2), (2,2), (0,2)\}$

    (b) $\{(0,0), (1,1), (1,2), (2,1), (0,2)\}$

    (c) $\{(0,0), (0,1), (0,2), (1,0), (2,0)\}$

    (d) $\{(0,1), (1,2), (0,2)\}$

    (e) $\{(0,0), (1,1), (2,2), (1,2), (2,1)\}$

2. Suppose $X = \{0, 1, 2, 3\}, Y = \{4, 5, 6\}$ and $Z = \{7, 8\}$. Further suppose

    - $R :\in X \leftrightarrow Y$ is equal to $\{(0,4), (0,5), (1,5), (1,6), (3,4)\}$,
    - $S :\in X \leftrightarrow Z$ is equal to $\{(0,7), (1,7), (2,7), (3,7), (2,8)\}$,
    - $T :\in Y \leftrightarrow Z$ is equal to $\{(4,7), (4,8), (5,7)\}$,
    - $U :\in Y \leftrightarrow X$ is equal to $\{(4,2), (5,2), (6,2)\}$ and
    - $V :\in Z \leftrightarrow Z$ is equal to $\{(7,8)\}$.

Decide whether each of the following relations can be described as reflexive, symmetric, antisymmetric, transitive, or a combination of these. Which are equivalence relations, and what are the corresponding equivalence classes?

(a) I_X

(b) $R \mathbin{\overset{\circ}{\scriptstyle\circ}} R^\sim$

(c) $U \mathbin{\overset{\circ}{\scriptstyle\circ}} U^\sim$

(d) $T \mathbin{\overset{\circ}{\scriptstyle\circ}} T^\sim$

(e) $S \mathbin{\overset{\circ}{\scriptstyle\circ}} S^\sim$

(f) $U^\sim \mathbin{\overset{\circ}{\scriptstyle\circ}} U$

(g) $R^\sim \mathbin{\overset{\circ}{\scriptstyle\circ}} R$

(h) $T^\sim \mathbin{\overset{\circ}{\scriptstyle\circ}} T$

(i) $S^\sim \mathbin{\overset{\circ}{\scriptstyle\circ}} S$

(j) V

(k) $(R \mathbin{\overset{\circ}{\scriptstyle\circ}} R^\sim) \cup I_X$

(l) $(R \mathbin{\overset{\circ}{\scriptstyle\circ}} R^\sim) \mathbin{\overset{\circ}{\scriptstyle\circ}} (R \mathbin{\overset{\circ}{\scriptstyle\circ}} R^\sim)$

(m) $(R \mathbin{\overset{\circ}{\scriptstyle\circ}} R^\sim) \mathbin{\overset{\circ}{\scriptstyle\circ}} (R \mathbin{\overset{\circ}{\scriptstyle\circ}} R^\sim) \cup I_X$

(n) $V \cap V^\sim$

3. How many equivalence relations on a set A are also functions?

4. Find the symmetric closures and transitive closures of each of the following relations on $\{0, 2, 4\}$:

(a) $\{(0, 0), (0, 2), (0, 4)\}$

(b) $\{(2, 0), (0, 2)\}$

(c) $\{(0, 0), (2, 4)\}$

(d) $\{(2, 4), (4, 2), (2, 0)\}$

CHAPTER 7

Functions

7.1 Introduction

In this chapter we shall be looking at how functions can be modelled using sets. We shall start however with an intuitive definition of the concept of function and consider the ways in which functions can be manipulated from this point of view. Subsequently we shall model functions by sets of ordered pairs; this enables us to give a more precise 'mathematical' definition of a function, and clearer definitions for the various operations on functions.

Note that in the intuitive view, you will probably find some of the concepts vague and hard to comprehend; this is often the case when we adopt an intuitive, non-mathematical approach to any concept. The great power of mathematics is its ability to make ideas clear and unambiguous; sadly this is still largely ignored by the software community. You will subsequently find that the use of sets to model functions makes the notions much clearer – always assuming that you have taken the trouble to understand the material in the earlier chapters!

7.2 An intuitive view of functions

First it is necessary to introduce a few intuitive 'definitions' and some notation for functions.

Definition 7.1 A function is a procedure which gives a *unique* output for any suitable input. The set of suitable inputs is called the *domain* of the function while the set of outputs which are possible is called the *range* of the function.

For example, we may define the function *SquareRoot* such that the square of the output is equal to the input. The domain and range could each be specified as the set of real numbers greater than or equal to 0, $\{ x :\in \mathbb{R} \mid x \geqslant 0 \}$. Thus *SquareRoot* will take an input of 4 and return an output of 2, since $2^2 = 4$, 4 is in the domain, and 2 is in the range. An input of -4 will result in an error since -4 is not in the domain.

Fact 7.2 A function can be given an identifying label. Often in mathematics this label is a single letter such as *f* or *g* or *h*, but longer names are possible such as *SquareRoot* or *Cube*.

Fact 7.3 A function is not totally defined unless its domain and range are stipulated.

As another example, suppose we define *Cube* to take any integer as input and output its cube. The domain is the set of integers, $\mathbb{Z}$. Although the range has not been stipulated explicitly, it can be deduced from the definition, and is also equal to $\mathbb{Z}$.

Notation 7.4 The domain of a function *f* is denoted $\mathrm{dom}\,f$, while the range of a function *f* is denoted $\mathrm{ran}\,f$.

Examples 7.5 Evaluate each of the following:

1. dom *SquareRoot*
2. ran *SquareRoot*
3. dom *Cube*
4. ran *Cube*

Solution 7.6

1. $\mathrm{dom}\,SquareRoot = \{\, x :\in \mathbb{R} \mid x \geqslant 0 \,\}$
2. $\mathrm{ran}\,SquareRoot = \{\, x :\in \mathbb{R} \mid x \geqslant 0 \,\}$
3. $\mathrm{dom}\,Cube = \mathbb{Z}$
4. $\mathrm{ran}\,Cube = \mathbb{Z}$

Definition 7.7 If *x* is in the domain of a function *f*, i.e. $x \in \mathrm{dom}\,f$, then we can input *x* to the function to obtain an output, *y*. We say that the function is *applied* to *x*, and that *y* is the *result* of this application.

Notation 7.8 The application of a function *f* to *x* can be denoted in several different ways, including:

- $f\,x$
- $x.f$
- $f(x)$

Usually we shall use just the last of these notations.

The uniqueness of output from a function can now be expressed using function application:

$$\forall x :\in \mathrm{dom}\,f;\ y,z :\in \mathrm{ran}\,f \bullet f(x) = y \wedge f(x) = z \Rightarrow y = z \tag{7.1}$$

Note that a function cannot be applied to a value *x* which is not in its domain. Similarly the result *y* of a function application must always lie in the range of the function.

Definition 7.9 It is often convenient to imagine that the domain of a function is a subset of another set known as the *source*. Similarly, the range is a subset of another set called the *target* or *codomain*.

Examples 7.10 Suggest suitable source and target sets for

1. *SquareRoot*
2. *Cube*

Solution 7.11

1. Usually we choose the source set of *SquareRoot* to be the set of real numbers, $\mathbb{R}$. Only those values which are in the domain, $\{x :\in \mathbb{R} \mid x \geqslant 0\}$, will give a meaningful answer. For example with an electronic calculator it is possible to enter any number then press the $\boxed{\sqrt{}}$ function key; for negative numbers, however, most calculators will generate an 'error' message. On more advanced calculators, it is true that an answer is obtained – in fact a 'complex' number – but the $\boxed{\sqrt{}}$ key in this case does not correspond to our *SquareRoot* function. The function which does correspond to the more advanced $\boxed{\sqrt{}}$ key, *ComplexSquareRoot* say, has different source and target sets to the *SquareRoot* function; the source and target sets of *ComplexSquareRoot* are both equal to the set of so-called complex numbers.
2. The source and target sets of *Cube* could be chosen to be $\mathbb{Z}$ for both. In that case the function could be applied to any element of the source set.

Definition 7.12 We see that for some functions the domain is equal to the source set; such a function is said to be *total*.

Definition 7.13 One special function is the *identity* function; the identity function gives an output value equal to the input value.

Definition 7.14 For a given source set X, the identity function, I, is such that I can be applied to any element x of the source X to give a result equal to x:

$$\forall x :\in X \bullet I(x) = x$$

is T.

For example if $X = \{1, 2, 3\}$ then the identity function, I, is such that $I(1) = 1$, $I(2) = 2$ and $I(3) = 3$.

In fact there is a different identity function for each possible source set.

Fact 7.15 All identity functions are total.

Notation 7.16 The identity function with source X will be denoted by I_X.

This notation is the same as that used for the identity relation on X; later we shall see that the identity relation and the identity function can be modelled by the same set of ordered pairs.

7.2.1   Operations on functions

Composition

Suppose that f and g are two functions; then the output from f can be input to g provided it lies in the domain of g. That is

$$\operatorname{ran} f \subseteq \operatorname{dom} g$$

For example suppose that f and g both have $\mathbb{N}$ as domain, and that f adds 1 to an input integer, while g multiplies an input integer by 5; f and g are chosen such that the following is true:

$$\forall x :\in \mathbb{N} \bullet f(x) = x + 1 \wedge g(x) = 5 * x$$

Then applying f to 3 gives $f(3) = 4$ which can be input to g to give $g(4) = 5 * 4 = 20$. That is we can write

$$g(f(3)) = g(4) = 20$$

Clearly we have defined a third function h such that

$$\forall x :\in \mathbb{N} \bullet h(x) = g(f(x))$$

is true. We say that h is equal to the *functional composition* of g and f.

Notation 7.17 The composition of g and f is denoted by $g \circ f$. Thus $(g \circ f)(x)$, that is $g \circ f$ applied to x, is equivalent to $g(f(x))$.

Hence for f and g as defined above, $(g \circ f)(3) = 20$.

What happens when the output from f does not lie in the domain of g? This problem frequently occurs in using electronic calculators, and suggests that we should still be able to talk of the composition of two functions even when the range of the first function we apply is different to the domain of the second function. For example suppose we are doing a calculation in which we add 1 then press the $\boxed{\sqrt{}}$ key. We can think of this calculation in terms of two functions f and g which have source and target sets equal to $\mathbb{Z}$. The function f has domain equal to $\mathbb{Z}$ and subtracts 1; g has domain equal to $\mathbb{N}$ and takes the positive square root. Then using -3 as the starting point for our calculation, we have $g(f(-3)) = \sqrt{(-4)}$, which cannot be evaluated because $f(-3) = -4$ does not lie within the domain of g. The domain of $g \circ f$ is in fact equal to $\mathbb{N}_1$, the set of integers greater than or equal to 1. Thus although f is a total function, neither g nor $g \circ f$ are.

Definition 7.18 The functional composition of two functions g and f is obtained by using the output from f as input to g, provided that the target set of f is the same as the source set of g.

If the source set of g is not the same as the target set of f, then the functional composition is undefined.

Fact 7.19 In general, for any two functions f and g

$$\mathrm{dom}(g \circ f) \subseteq \mathrm{dom} f$$

An interesting situation arises when one of the two functions is an identity function.

Fact 7.20 If f is a function with source A and codomain B then

$$I_B \circ f \; = \; f \tag{7.2}$$
$$f \circ I_A \; = \; f \tag{7.3}$$

Examples 7.21 If

$$\forall x :\in \mathbb{Z} \bullet f(x) = x - 1$$

where the domain of f is $\mathbb{Z}$, and

$$\forall x :\in \mathbb{N} \bullet g(x) = 6 * x$$

where the domain of g is $\mathbb{N}$, find

1. $(g \circ f)(2)$
2. $(f \circ g)(2)$

Solution 7.22

1. $(g \circ f)(2) = g(f(2)) = g(2 - 1) = g(1) = 6 * 1 = 6$
2. $(f \circ g)(2) = f(g(2)) = f(6 * 2) = f(12) = 12 - 1 = 11$

Note that the two answers are not equal.

Fact 7.23 In general $g \circ f \neq f \circ g$.

7.2.2   Inverse function

Definition 7.24 Suppose that we have a pair of functions f and g such that

$$\mathrm{dom}\, g = \mathrm{ran} f$$
$$\mathrm{ran}\, g = \mathrm{dom} f$$

and that $y = f(x)$ is true if and only if $x = g(y)$ is true. Then f and g are said to be *inverse functions* of each other.

For example suppose *AddOne* adds 1 to any integer value, while *SubtractOne* subtracts 1 from any integer value. Then *AddOne*(3) = 4 while *SubtractOne*(4) = 3; *SubtractOne*(−1) = −2 while *AddOne*(−2) = −1.

Notation 7.25 The inverse function to a function f is denoted by f^{-1}.

So we can write $AddOne^{-1} = SubtractOne$ and $SubtractOne^{-1} = AddOne$.

The effect of one function is opposite to that of its inverse function; in particular, passing the output from a function f as input to the inverse function f^{-1} results in an output from f^{-1} equal to the input to f.

Fact 7.26 If a function f has an inverse function f^{-1}, then for any x in the domain of f it is true that $(f^{-1} \circ f)(x) = x$.

Note that not every function has an inverse function. Suppose that we have a function *Square* for which $Square(-2) = Square(2) = 4$, then applying an inverse function to 4 would need to result in a value which was equal to both 2 and −2; clearly this cannot be possible, and so the function *Square* could never have an inverse function.

7.2.3   Functional override

Given a function g with domain equal to some set A, we may want to define a new function which acts just like g for inputs taken from A, but like some other function f for inputs not in A (but still in the domain of f). Such a function is called *override* of f by g.

Notation 7.27 The override of f by g is denoted as $f \oplus g$.

Examples 7.28 If $\forall x :\in \mathbb{Z} \bullet f(x) = x - 1$ where the domain of f is $\mathbb{Z}$, and $\forall x :\in \mathbb{N} \bullet g(x) = 6 * x$ where the domain of g is $\mathbb{N}$, find

1. $f \oplus g(2)$
2. $f \oplus g(-1)$
3. $g \oplus f(2)$
4. $g \oplus f(-1)$

Solution 7.29

1. Since $2 \in \text{dom } g$, the value of $f \oplus g(2)$ is determined by g itself:

$$f \oplus g(2) = g(2) = 12$$

2. Since $-1 \notin \text{dom } g$ but $-1 \in \text{dom } f$ the value of $f \oplus g(-1)$ is determined by f:

$$f \oplus g(-1) = f(-1) = -2$$

3. $g \oplus f(2) = f(2) = 1$
4. $g \oplus f(-1) = f(-1) = -2$

Note that $g \oplus f = f$.

Fact 7.30 If $\text{dom } g \subseteq \text{dom } f$ then $g \oplus f = f$.

7.3   Modelling functions

7.3.1   Modelling functions by diagrams

As we have already seen in the chapters on relations, diagrams are frequently used in modelling. Two particular kinds of diagram used for functions (as well as relations) are:

- Cartesian plots – Input and output values are plotted as pairs of Cartesian coordinates on an '*x–y* graph', with input values on the *x*-axis. This should be familiar to you from elementary mathematics; an example is given in Figure 7.1. The source set is represented by the horizontal axis, the target set by the vertical axis.

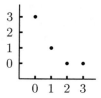

Figure 7.1 *An example of a Cartesian plot.*

- Arrow diagrams – The source and target sets are represented by two lists (usually both vertical); each input value in the first list is joined to the corresponding output value in the second list by an arrow. Figure 7.2 gives an example.

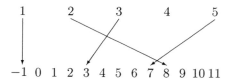

Figure 7.2 *An example of an arrow diagram.*

7.3.2   Modelling functions by product sets

A function can be modelled by the set of pairs of coordinates, as for example used in plotting the function. In fact the two diagrams given above both refer to the function f with domain given by $\operatorname{dom} f = \{1, 2, 3, 4\}$ and for which

$$\forall x :\in \operatorname{dom} f \bullet f(x) = x * (x - 2)$$

so that the following set of coordinate pairs can be calculated:

$$\{(1,-1),(2,0),(3,3),(4,8)\}$$

Examples 7.31 For each of the following functions, find the set of ordered pairs which model the function, and use this set to represent the function as a Cartesian plot and as an arrow diagram:

1. $\mathrm{dom}\,g = \{0,1,2\}$ with $\forall x :\in \mathrm{dom}\,g \bullet g(x) = x$
2. $\mathrm{dom}\,h = \{0,2,4\}$ with $\forall x :\in \mathrm{dom}\,h \bullet h(x) = 5 - x$

Solution 7.32

1. The set of ordered pairs is $\{(0,0),(1,1),(2,2)\}$ with diagrams as shown in Figure 7.3:

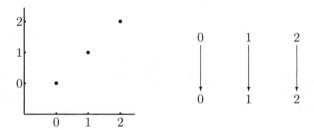

Figure 7.3 $g(x) = x$.

2. The set of ordered pairs is $\{(0,5),(2,3),(4,1)\}$ with diagrams as shown in Figure 7.4:

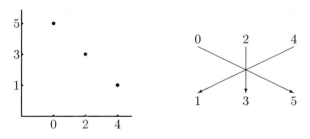

Figure 7.4 $h(x) = 5 - x$.

In discrete mathematics we often talk about the set which models a function as actually being the function itself. This is a very common and useful convention which is generally adopted throughout this book. In particular it is usual to equate a function to the set which models that function. For example, for the function f with $\operatorname{dom} f = \{1, 2, 3, 4\}$ and

$$\forall x :\in \operatorname{dom} f \bullet f(x) = x * (x - 2)$$

we often write

$$f = \{(1, -1), (2, 0), (3, 3), (4, 8)\}$$

This is similar to the way in which we often regard a relation as being the set of ordered pairs which models it. In fact, we can now regard a 'function' as a special kind of 'relation'.

Where it is desired to emphasize the function rather than its model, we shall use the term *operation*.

The 'application of a function f to x' is regarded as an operation on the set f of ordered pairs and x, to give the value $f(x)$.

Examples 7.33 If $f = \{(1, 2), (2, 6), (3, 2), (4, 2)\}$, what is the value of $f(2)$?

Solution 7.34 $f(2)$ is the result of applying $\{(1, 2), (2, 6), (3, 2), (4, 2)\}$ to the value 2; the application operation involves passing through the first coordinates of the set until a match is obtained with 2; the result of the function application is read off as the second coordinate. In this case the answer obtained is 6.

It is important to realize that although a function from elements in a source set A to a target set B can be modelled by a subset of $A \times B$, not all subsets of $A \times B$ can be models of functions; if the set of ordered pairs contains two elements (x, y) and (x, z) in which y and z are different, then there would be two possible outputs corresponding to an input x.

Examples 7.35 Which of the following 'relations' are 'functions' – that is to say, which sets of ordered pairs could be used to model functions?

1. $\{(1, 2), (2, 3), (3, 1)\}$
2. $\{(1, 2), (2, 2), (3, 2)\}$
3. $\{(1, 2), (2, 3), (1, 1)\}$
4. $\{(1, 2), (2, 2), (3, 2), (4, 1), (5, 3)\}$
5. $\{(1, 2)\}$
6. $\varnothing$

Solution 7.36 Intuitively we can check whether a set is a function by seeing how many times each first coordinate occurs; if there are no repetitions, then the set is indeed a function.

1. $\{(1, 2), (2, 3), (3, 1)\}$ is a function since the first coordinates $1, 2, 3$ are each paired with just one value.

2. $\{(1,2),(2,2),(3,2)\}$ is a function since $1,2,3$ each occur just once in the first coordinates; note that 2 occurs three times in the second coordinates, but this does not alter the fact that we have a function.

3. In $\{(1,2),(2,3),(1,1)\}$ there are two ordered pairs $(1,1)$ and $(1,2)$ in which 1 occurs as the first coordinates. This set is *not* a function.

4. $\{(1,2),(2,2),(3,2),(4,1),(5,3)\}$ is a function.

5. $\{(1,2)\}$ is also a function, although it maps just one value 1 to the value of 2.

6. $\varnothing$ is also a function since there are no repeated first coordinates; indeed there are no first coordinates whatsoever.

This last example with the empty set poses something of a difficulty; with our definition of a function expressed in words, it may be not quite clear whether or not $\varnothing$ should count as a function. This is one of the dangers of using an informal definition of a mathematical object. It is therefore usually preferable to give a *formal* definition using predicate logic in order to avoid ambiguities or questionable cases. The following definition is derived from definition 7.1.

Definition 7.37 A subset f of $A \times B$ is a function if

$$\forall x :\in A;\ y, z :\in B \bullet ((x,y) \in f \wedge (x,z) \in f) \Rightarrow y = z$$

is T.

One way of looking at this definition is to suppose that we have two different programs, one of which generates a value x, the other a value y. If whenever (x, y) and (x, z) are both in the set f it is true that $y = z$, then f is a function. This rather cumbersome logical definition provides us with a 'mechanistic' (if somewhat impractical) way of determining whether f is a function for the cases in which f is finite.

Examples 7.38 If $A = \{0, 1, 2\}$ and $B = \{0, 1\}$ determine whether f in each case below is a function or not:

1. $f = \{(1, 1), (0, 1)\}$
2. $f = \{\}$

Solution 7.39

1. In this case a quick inspection will tell us that f is indeed a function. However, we can use the formal definition to build a table of all the possible combinations of x, y and z. For $f = \{(1, 1), (0, 1)\}$ we have

x	y	z	$((x,y) \in f \land (x,z) \in f) \Rightarrow y = z$	TruthValue
0	0	0	$((0,0) \in f \land (0,0) \in f) \Rightarrow 0 = 0$	$F \Rightarrow T \equiv T$
0	0	1	$((0,0) \in f \land (0,1) \in f) \Rightarrow 0 = 1$	$F \Rightarrow F \equiv T$
0	1	0	$((0,1) \in f \land (0,0) \in f) \Rightarrow 1 = 0$	$F \Rightarrow F \equiv T$
0	1	1	$((0,1) \in f \land (0,1) \in f) \Rightarrow 1 = 1$	$T \Rightarrow T \equiv T$
1	0	0	$((1,0) \in f \land (1,0) \in f) \Rightarrow 0 = 0$	$F \Rightarrow T \equiv T$
1	0	1	$((1,0) \in f \land (1,1) \in f) \Rightarrow 0 = 1$	$F \Rightarrow F \equiv T$
1	1	0	$((1,1) \in f \land (0,0) \in f) \Rightarrow 1 = 0$	$F \Rightarrow F \equiv T$
1	1	1	$((1,1) \in f \land (1,1) \in f) \Rightarrow 1 = 1$	$T \Rightarrow T \equiv T$
2	0	0	$((2,0) \in f \land (2,0) \in f) \Rightarrow 0 = 0$	$F \Rightarrow T \equiv T$
2	0	1	$((2,0) \in f \land (2,1) \in f) \Rightarrow 0 = 1$	$F \Rightarrow F \equiv T$
2	1	0	$((2,1) \in f \land (2,0) \in f) \Rightarrow 1 = 0$	$F \Rightarrow F \equiv T$
2	1	1	$((2,1) \in f \land (2,1) \in f) \Rightarrow 1 = 1$	$F \Rightarrow T \equiv T$

Hence we see that $\forall x :\in A; \ y, z :\in B \bullet (x,y) \in f \land (x,z) \in f \Rightarrow y = z$ is true, and conclude that f is a function.

2. For the case when $f = \{\}$ common sense can lead us astray, or perhaps leave us confused. Is the empty set a function or not? Using the formal definition removes any chance of ambiguity in the concept of a function.

x	y	z	$((x,y) \in f \land (x,z) \in f) \Rightarrow y = z$	TruthValue
0	0	0	$((0,0) \in f \land (0,0) \in f) \Rightarrow 0 = 0$	$F \Rightarrow T \equiv T$
0	0	1	$((0,0) \in f \land (0,1) \in f) \Rightarrow 0 = 1$	$F \Rightarrow F \equiv T$
0	1	0	$((0,1) \in f \land (0,0) \in f) \Rightarrow 1 = 0$	$F \Rightarrow F \equiv T$
0	1	1	$((0,1) \in f \land (0,1) \in f) \Rightarrow 1 = 1$	$F \Rightarrow T \equiv T$
1	0	0	$((1,0) \in f \land (1,0) \in f) \Rightarrow 0 = 0$	$F \Rightarrow T \equiv T$
1	0	1	$((1,0) \in f \land (1,1) \in f) \Rightarrow 0 = 1$	$F \Rightarrow F \equiv T$
1	1	0	$((1,1) \in f \land (0,0) \in f) \Rightarrow 1 = 0$	$F \Rightarrow F \equiv T$
1	1	1	$((1,1) \in f \land (1,1) \in f) \Rightarrow 1 = 1$	$F \Rightarrow T \equiv T$
2	0	0	$((2,0) \in f \land (2,0) \in f) \Rightarrow 0 = 0$	$F \Rightarrow T \equiv T$
2	0	1	$((2,0) \in f \land (2,1) \in f) \Rightarrow 0 = 1$	$F \Rightarrow F \equiv T$
2	1	0	$((2,1) \in f \land (2,0) \in f) \Rightarrow 1 = 0$	$F \Rightarrow F \equiv T$
2	1	1	$((2,1) \in f \land (2,1) \in f) \Rightarrow 1 = 1$	$F \Rightarrow T \equiv T$

Hence we see that $\forall x :\in A; \ y, z :\in B \bullet (x,y) \in f \land (x,z) \in f \Rightarrow y = z$ is true; $f = \varnothing$ is a function.

Notation

A function is associated with the notion of passing from an input value x to an output value y. To remind us of this connection, we often use arrow symbols when dealing with functions; several different sorts of arrow are used and care must be taken not to confuse them.

Notation 7.40 Ordered pairs are often represented using 'maplet' notation, that is (x, y) is written as $x \mapsto y$.

Fact 7.41 When an ordered pair is represented using maplet notation, it is usually referred to as a *maplet*.

Notation 7.42 The set of functions with domain A and codomain B is denoted by $A \to B$.

Thus $f :\in A \to B$ declares f to be a function with $\operatorname{dom} f = A$ and $\operatorname{ran} f \subseteq B$.

Examples 7.43 If $X = \{3, 4, 5\}$ and $Y = \{1, 2\}$ evaluate $X \to Y$.

Solution 7.44 Each element of X is mapped to either 1 or 2:

$$
\begin{aligned}
X \to Y = \{\ &\{(3,1), (4,1), (5,1)\} \\
&\{(3,1), (4,1), (5,2)\} \\
&\{(3,1), (4,2), (5,1)\} \\
&\{(3,1), (4,2), (5,2)\} \\
&\{(3,2), (4,1), (5,1)\} \\
&\{(3,2), (4,1), (5,2)\} \\
&\{(3,2), (4,2), (5,1)\} \\
&\{(3,2), (4,2), (5,2)\}\}
\end{aligned}
$$

Using maplet notation this could be written:

$$
\begin{aligned}
X \to Y = \{\ &\{3 \mapsto 1, 4 \mapsto 1, 5 \mapsto 1\} \\
&\{3 \mapsto 1, 4 \mapsto 1, 5 \mapsto 2\} \\
&\{3 \mapsto 1, 4 \mapsto 2, 5 \mapsto 1\} \\
&\{3 \mapsto 1, 4 \mapsto 2, 5 \mapsto 2\} \\
&\{3 \mapsto 2, 4 \mapsto 1, 5 \mapsto 1\} \\
&\{3 \mapsto 2, 4 \mapsto 1, 5 \mapsto 2\} \\
&\{3 \mapsto 2, 4 \mapsto 2, 5 \mapsto 1\} \\
&\{3 \mapsto 2, 4 \mapsto 2, 5 \mapsto 2\}\}
\end{aligned}
$$

Sometimes we do not wish to state precisely what the domain of a function is, but simply to state that it is a subset of some source set, for example A.

Notation 7.45 The set of all functions whose source set is A and whose target (or codomain) is B is denoted by $f :\in A \nrightarrow B$.

Thus if $f :\in A \nrightarrow B$, then $\operatorname{dom} f \subseteq A$. Note that it may well be that $\operatorname{dom} f$ is indeed the same as the source set A.

Fact 7.46 $A \to B \subseteq A \nrightarrow B$

Examples 7.47 If $X = \{3, 4\}$ and $Y = \{1, 2\}$ evaluate $X \nrightarrow Y$.

Solution 7.48 Each element of X is mapped to either 1 or 2, or is not mapped to anything:

$$X \nrightarrow Y = \{ \quad \{3 \mapsto 1, 4 \mapsto 1\}$$
$$\{3 \mapsto 1, 4 \mapsto 2\}$$
$$\{3 \mapsto 1\}$$
$$\{3 \mapsto 2, 4 \mapsto 1\}$$
$$\{3 \mapsto 2, 4 \mapsto 2\}$$
$$\{3 \mapsto 2\}$$
$$\{4 \mapsto 1\}$$
$$\{4 \mapsto 2\}$$
$$\{\}\}$$

Note that $\varnothing \in X \nrightarrow Y$ and also that $X \rightarrow Y \subseteq X \nrightarrow Y$.

7.4   Modelling operations on functions

There are several operations which can be applied to functions:

- $\operatorname{dom} f$ gives the domain of f;
- $\operatorname{ran} f$ gives the range of f;
- $g \circ f$ gives the composition of g and f;
- f^{-1} provided the inverse function exists;
- $f \oplus g$ gives the override of f by g.

It is possible to model these operations in terms of sets.

7.4.1   General set operations

Because we regard a function as a set of ordered pairs, it is of course possible to apply all set operations to functions.

Examples 7.49 Suppose we have two sets of ordered pairs, $f = \{2 \mapsto 4, 3 \mapsto 5\}$ and $g = \{2 \mapsto 1, 3 \mapsto 5, 4 \mapsto 0\}$, which are used to model functions with source and target sets of $\mathbb{Z}$. Evaluate each of the following sets, and decide which results could model functions.

1. $f \cup g$
2. $f \cap g$
3. $\mathbb{P} f$
4. $f \times (f \cap g)$

Solution 7.50 To help remind myself that some of the answers may not be functions, I prefer *not* to use the maplet notation. Strictly speaking, there is nothing wrong with the maplet notation even when we do not have a function since the notation is merely an alternative representation for ordered pairs, and the word 'maplet' means the same as 'ordered pairs'. However, I find restricting the use of maplet notation to *known* functions helpful. Some people prefer to use the maplet notation all the time, while others never use it.

1. $f \cup g = \{(2,4),(2,1),(3,5),(4,0)\}$ which is not a function since both $(2,4)$ and $(2,1)$ are elements of the set.
2. $f \cap g = \{(3,5)\}$ which is a function. This result can be *generalized*: no matter what functions are chosen for f and g, the result of $f \cap g$ will also be a function.
3. $\mathbb{P}f = \{\{(2,4),(3,5)\},\{(3,5)\},\{(2,4)\},\varnothing\}$ which cannot be a function since the elements are themselves sets, and not ordered pairs (maplets).
4. $f \times (f \cap g) = \{((2,4),(3,5)),((3,5),(3,5))\}$ which is a set of ordered pairs of ordered pairs. This is indeed a function, and we can, for example, write

$$f \times (f \cap g)(2,4) = (3,5)$$

In general, however, $f \times (f \cap g)$ will *not* be a function.

Note that in the last example we have used the symbol $\times$ to indicate the Cartesian product of two sets. Mathematicians commonly define another operation known as the *cross product of two functions* which always results in a function; confusingly the symbol $*$ is used to indicate this operation.

7.4.2   Domain

Definition 7.51 The domain, $\operatorname{dom} f$, of the set corresponding to a function f is the set of all the first coordinates.

Examples 7.52 If $f = \{1 \mapsto 3, 4 \mapsto 2, 5 \mapsto 3\}$ find $\operatorname{dom} f$.

Solution 7.53 In the set enumeration for f, the first maplet has first coordinate 1; the second maplet has first coordinate 4; the third maplet has first coordinate 5.

$$\operatorname{dom} f = \{1,4,5\}$$

Notice that it is irrelevant that both 1 and 5 map to the same value 3.

7.4.3   Range

Definition 7.54 The range, $\operatorname{ran} f$, of the set corresponding to a function f is the set of all the second coordinates.

Examples 7.55 If $f = \{1 \mapsto 3, 4 \mapsto 2, 5 \mapsto 3\}$ find $\operatorname{ran} f$.

Solution 7.56 In the set enumeration for f, the first maplet has second coordinate 3; the second maplet has second coordinate 2; the third maplet has second coordinate 3.

$$\operatorname{ran} f = \{3, 2, 3\} = \{2, 3\}$$

Notice that since both 1 and 5 map to the same value 3, this value needs only to be represented once in the set enumeration for $\operatorname{ran} f$.

7.4.4   Composition

The composition of two functions can be modelled using sets. Note that the composition of two functions always results in a function.

Suppose $f = \{-2 \mapsto 4, -1 \mapsto 1, 0 \mapsto 0, 1 \mapsto 1, 2 \mapsto 4\}$, and $g = \{-1 \mapsto -2, 0 \mapsto 1, 1 \mapsto 2\}$ then we can find the values of $g \circ f$ and $f \circ g$ to be $\{-1 \mapsto 2, 0 \mapsto 1, 1 \mapsto 2\}$ and $\{-1 \mapsto 4, 0 \mapsto 1, 1 \mapsto 4\}$ respectively.

The calculation of these results is perhaps most easily achieved as a diagram. For example in Figure 7.5, the calculation of $g \circ f$ is shown. Each function is modelled as an arrow diagram with the domain of g and the range of f in a common (middle) box. Each element of $g \circ f$ corresponds to a complete path from the first box to the third box.

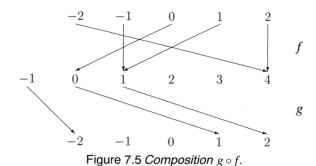

Figure 7.5 *Composition $g \circ f$.*

It can be seen that in terms of sets, functional composition is very similar to relational composition. The only real difference is in the order in which the sets are given: since functions may be regarded as special kinds of relations, then the functional composition $g \circ f$ could equally well be expressed using relational composition as $f \,\substack{\circ \\ 9}\, g$.

Fact 7.57

$$f \,\substack{\circ \\ 9}\, g \mathrel{\widehat{=}} g \circ f$$

So for the example above in which $f = \{-2 \mapsto 4, -1 \mapsto 1, 0 \mapsto 0, 1 \mapsto 1, 2 \mapsto 4\}$, and $g = \{-1 \mapsto -2, 0 \mapsto 1, 1 \mapsto 2\}$ we *could* write

$$f \,\mathring{,}\, g \;=\; \{-1 \mapsto 2, 0 \mapsto 1, 1 \mapsto 2\}$$
$$g \,\mathring{,}\, f \;=\; \{-1 \mapsto 4, 0 \mapsto 1, 1 \mapsto 4\}$$

We can apply these functions to values, for example:

$$f \,\mathring{,}\, g(-1) \;=\; g(f(-1)) = 2 \tag{7.4}$$
$$g \,\mathring{,}\, f(-1) \;=\; f(g(-1)) = 4 \tag{7.5}$$

Notice that the notation with '$\circ$' seems a more natural notation when applying the composite function to a value:

$$g \circ f(-1) \;=\; g(f(-1)) = 2 \tag{7.6}$$
$$f \circ g(-1) \;=\; f(g(-1)) = 4 \tag{7.7}$$

In general therefore we shall not use the '$\mathring{,}$' symbol for composition of functions, but restrict its use to modelling the composition of relations.

7.4.5   Inverse functions

Where an inverse function f^{-1} exists it is obtained by reversing the order of the coordinates in each orderd pair. For example if $f = \{(1,5), (2,6), (3,4)\}$ then we can write down $f^{-1} = \{(5,1), (6,2), (4,3)\}$. In maplet notation we have $f = \{1 \mapsto 5, 2 \mapsto 6, 3 \mapsto 4\}$ then we can write down $f^{-1} = \{5 \mapsto 1, 6 \mapsto 2, 4 \mapsto 3\}$.

We can check this result by calculation.

Examples 7.58 If $f = \{1 \mapsto 5, 2 \mapsto 6, 3 \mapsto 4\}$ and $f^{-1} = \{5 \mapsto 1, 6 \mapsto 2, 4 \mapsto 3\}$, calculate

1. $f^{-1} \circ f$
2. $f \circ f^{-1}$

Solution 7.59

1. $f^{-1} \circ f = \{1 \mapsto 1, 2 \mapsto 2, 3 \mapsto 3\}$ (see Figure 7.6).
2. $f \circ f^{-1} = \{5 \mapsto 5, 6 \mapsto 6, 4 \mapsto 4\}$ (see Figure 7.7).

Thus we obtain a subset of the identity function in each case.

However if $f = \{(1,5), (2,5), (3,4)\}$ then reversing the coordinates gives $\{(5,1), (5,2), (4,3)\}$, which is not a function (because 5 appears as first coordinate paired with two different values).

Fact 7.60 If we regard a function as a special kind of relation, then we can apply relational inverse (denoted by $\sim$) to obtain another relation. In general this relation will not be a function.

Thus for $f = \{(1,5), (2,5), (3,4)\}$ we may write $f^\sim = \{(5,1), (5,2), (4,3)\}$ even though this does not represent a function.

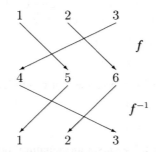

Figure 7.6 *Composition $f^{-1} \circ f$.*

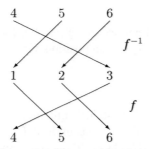

Figure 7.7 *Composition $f \circ f^{-1}$.*

7.4.6 Function override

Suppose that $f, g :\in \mathbb{Z} \nrightarrow \mathbb{Z}$ with

$$f = \{-2 \mapsto -3, -1 \mapsto 0, 0 \mapsto 1, 2 \mapsto 3\}$$
$$g = \{-1 \mapsto -2, 0 \mapsto 1, 1 \mapsto 2\}$$

then we require

$$f \oplus g = \{-1 \mapsto -2, 0 \mapsto 1, 1 \mapsto 2, -2 \mapsto -3, 2 \mapsto 3\}$$
$$g \oplus f = \{-2 \mapsto -3, -1 \mapsto 0, 0 \mapsto 1, 2 \mapsto 3, 1 \mapsto 2\}$$

One way of approaching function override of f by g is to start with the set union, $f \cup g$; as we saw above, this does not always result in a function since there may be elements in both domains which map to different values. For example if $f = \{(1,3), (2,4)\}$ and $g = \{(3,5), (2,6)\}$ then

$$f \cup g = \{(1,3), (3,5), (2,4), (2,6)\}$$

To make a function from this it is necessary to lose either $(2,4)$ (from f) or $(2,6)$ (from g); the element from g overrides that from f, so $(2,4)$ is dropped while $(2,6)$ is retained.

We can express this more formally using set notation and predicate logic.

Definition 7.61

$$f \oplus g \stackrel{\frown}{=} f \cup g \setminus \{ x \mid x \in \mathrm{dom} f \wedge x \in \mathrm{dom}\, g \bullet x \mapsto f(x) \}$$

Several properties follow from this definition.

Fact 7.62 The override of two functions is always another function.

Fact 7.63

$$\mathrm{dom}(f \oplus g) = (\mathrm{dom} f) \cup (\mathrm{dom}\, g)$$

Examples 7.64 If $f, g :\in \mathbb{N} \nrightarrow \mathbb{N}$ is such that

$$f = \{ 2 \mapsto 4, 3 \mapsto 5 \}$$

and

$$g = \{ 2 \mapsto 1, 3 \mapsto 5, 4 \mapsto 0 \}$$

evaluate

1. $\{ x :\in \mathrm{dom}\, g \mid x \notin \mathrm{dom} f \bullet x \mapsto g(x) \}$
2. $\{ x :\in \mathrm{dom} f \mid x \notin \mathrm{dom}\, g \bullet x \mapsto f(x) \}$
3. $g \oplus f$
4. $f \oplus g$
5. $(f \oplus g) \oplus f$
6. $f \oplus (g \oplus f)$

Solution 7.65

1. $\{ x :\in \mathrm{dom}\, g \mid x \notin \mathrm{dom} f \bullet x \mapsto g(x) \} = \{ 4 \mapsto 0 \}$
2. $\{ x :\in \mathrm{dom} f \mid x \notin \mathrm{dom}\, g \bullet x \mapsto f(x) \} = \{\}$ because $\mathrm{dom} f \subseteq \mathrm{dom}\, g$
3. $g \oplus f = g \cup \{ 4 \mapsto 0 \} = \{ 2 \mapsto 4, 3 \mapsto 5, 4 \mapsto 0 \}$
4. $f \oplus g = g \cup \{\} = g$
5. $(f \oplus g) \oplus f = g \oplus f$
6. $f \oplus (g \oplus f) = f \oplus g$

The last two examples are particular examples of a general result; no matter what values are chosen for f and g, $(f \oplus g) \oplus f = g \oplus f$.

7.5   Special kinds of function

It is very useful to identify kinds of function with special properties; these are often used to model specific types of system or data structure. It is important to realize that several of these special properties depend upon the source and target sets; for such properties it is necessary to specify the source and target sets.

7.5.1 Total functions

Suppose that we have a function f with source set A and target set B. Then in general $\mathrm{dom} f \subseteq A$.

Definition 7.66 If $\mathrm{dom} f = A$ then f is said to be a *total* function on A.

This is illustrated in Figure 7.8 which represents the function

$$\{1 \mapsto 1, 2 \mapsto 1, 3 \mapsto 0, 4 \mapsto 3\}$$

in which the source set is $\{1, 2, 3\}$ and the target set is $\{0, 1, 2, 3\}$.

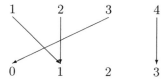

Figure 7.8 *A total function:* $\{1 \mapsto 1, 2 \mapsto 1, 3 \mapsto 0, 4 \mapsto 3\}$.

Some authors also introduce the term 'partial function' to mean *exactly* the same as our term 'function'.

Definition 7.67 If $\mathrm{dom} f \subseteq A$ then f is said to be a *partial* function on A.

This seems a rather strange thing to do, and can be highly misleading – according to the definition, all total functions are partial! Note carefully that 'partial' does not mean the same as 'not total'. The main justification for having the term is to emphasize that a function *may* or *may not* be total. It is recommended that a function is only described as 'partial' when it is considered necessary to emphasize that it may not be total; otherwise it should be described simply as a function.

Examples 7.68 Which of the following (partial) functions are total?

1. $\{red \mapsto 1, blue \mapsto 1, yellow \mapsto 2\}$ with source set $\{red, blue, yellow\}$ and target set $\{1, 2\}$.
2. $\{red \mapsto 1, blue \mapsto 1, yellow \mapsto 2\}$ with source set $\{red, blue, yellow, green\}$ and target set $\{1, 2\}$.
3. $\{red \mapsto 1, blue \mapsto 1, yellow \mapsto 2\}$ with source set $\{red, blue, yellow\}$ and target set $\{1, 2, 3\}$.

Solution 7.69

1. $\{red \mapsto 1, blue \mapsto 1, yellow \mapsto 2\}$ with source set $\{red, blue, yellow\}$ and target set $\{1, 2\}$ has domain equal to the source set; it is therefore total.

2. $\{red \mapsto 1, blue \mapsto 1, yellow \mapsto 2\}$ with source set $\{red, blue, yellow, green\}$ and target set $\{1, 2\}$ is not total since the element *green* of the source set is not in the domain.

3. $\{red \mapsto 1, blue \mapsto 1, yellow \mapsto 2\}$ with source set $\{red, blue, yellow\}$ and target set $\{1, 2, 3\}$ has domain equal to the source set and is therefore total. Note that the range does not contain all the elements of the target set; this has no bearing on whether the set is total however.

7.5.2 Surjections

We have just seen that a function whose domain is equal to the source set is called 'total'. We can also categorize a special type of function whose range is equal to its target set.

Definition 7.70 A *surjective* function is one whose range is equal to the whole of its target set.

This is illustrated in Figure 7.9 which represents the function $\{1 \mapsto 1, 2 \mapsto 1, 3 \mapsto 0\}$ in which the source set is $\{1, 2, 3\}$ and the target set is $\{0, 1\}$.

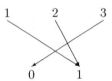

Figure 7.9 *A* surjective *function:* $\{1 \mapsto 1, 2 \mapsto 1, 3 \mapsto 0\}$.

Fact 7.71 A surjective function is sometimes referred to simply as a *surjection*.

Examples 7.72 Which of the following functions are surjective?

1. $\{1 \mapsto 5, 2 \mapsto 5, 3 \mapsto 6, 4 \mapsto 5\}$ with target set $\{5, 6, 7\}$
2. $\{1 \mapsto 5, 2 \mapsto 5, 3 \mapsto 6, 4 \mapsto 5\}$ with target set $\{5, 6\}$

Solution 7.73

1. $\mathrm{ran}\{1 \mapsto 5, 2 \mapsto 5, 3 \mapsto 6, 4 \mapsto 5\} = \{5, 6\}$. In this case the target set is $\{5, 6, 7\}$ and so we conclude that the given function is not surjective.
2. $\mathrm{ran}\{1 \mapsto 5, 2 \mapsto 5, 3 \mapsto 6, 4 \mapsto 5\} = \{5, 6\}$. In this case $\{5, 6\}$ equals the target set and so we conclude that the given function is surjective.

Note that in this case we *must* know the target set in order to decide whether a function is surjective.

Notation 7.74 The set of all (partial) surjections from source A to target (range) B is denoted by $A \twoheadrightarrow B$.

Notation 7.75 The set of all total surjections from source (domain) A to target (range) B is denoted by $A \twoheadrightarrow B$.

More formally we can write down the following definitions:

$$A \twoheadrightarrow B \ \widehat{=} \ \{f :\in A \twoheadrightarrow B \mid \mathrm{ran}\, f = B\}$$
$$A \twoheadrightarrow B \ \widehat{=} \ \{f :\in A \to B \mid \mathrm{ran}\, f = B\}$$

7.5.3 Injections

Definition 7.76 An *injective* function, or *injection*, is a function f with the special property that each element in the range of f is mapped to by a unique element in the domain of f.

$$\forall x, y :\in \ \mathrm{dom} f \bullet f(x) = f(y) \Rightarrow x = y$$

This idea can be seen more easily from Figure 7.10 which represents the function $\{1 \mapsto 2, 2 \mapsto 3, 3 \mapsto 0\}$; note that each element of the range has just one arrow pointing to it. Note also that, in general, the range of the function will be a subset of the target set while the domain will be a subset of the source set.

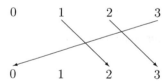

Figure 7.10 *An* injective *function:* $\{1 \mapsto 2, 2 \mapsto 3, 3 \mapsto 0\}$.

From these ideas we can write down the following practical definition.

Definition 7.77 A function f is injective if and only if $f^\sim$ is also a function.

Examples 7.78 Which of the following functions are injective?

1. $\{(1, 2), (2, 3)\}$
2. $\{(1, 2), (2, 3), (3, 2)\}$
3. $\varnothing$

Solution 7.79

1. $\{(1, 2), (2, 3)\}^\sim = \{(2, 1), (3, 2)\}$ which is also a function; the given function is injective.

2. $\{(1,2),(2,3),(3,2)\}^\sim = \{(2,1),(3,2),(2,3)\}$ which is not a function as the value 2 in the domain is mapped to two different values in the range, 1 and 3. The given function is not injective.

3. $\varnothing^\sim = \varnothing$ which is the null function again; $\varnothing$ is injective.

Note that we have not been given the source and target sets in these examples.

Fact 7.80 It is not necessary to know the source and target sets in order to decide whether a given function is injective.

Notation 7.81 The set of all (partial) injections from source A to target B is denoted by $A \rightarrowtail B$.

Notation 7.82 The set of all total injections from source (domain) A to target B is denoted by $A \rightarrowtail B$.

More formally we can write down the following definitions:

$$A \rightarrowtail B \;\; \widehat{=} \;\; \{f :\in A \twoheadrightarrow B \mid f^\sim \in B \twoheadrightarrow A\}$$
$$A \rightarrowtail B \;\; \widehat{=} \;\; \{f :\in A \to B \mid f^\sim \in B \twoheadrightarrow A\}\,.$$

7.6 Bijections

Definition 7.83 A *bijective* function or *bijection* is a function which is total, surjective and injective.

Notation 7.84 The set of all bijections from source A to target B is denoted by $A \rightarrowtail\!\!\!\twoheadrightarrow B$.

More formally we can write down the following definition:

$$A \rightarrowtail\!\!\!\twoheadrightarrow B \,\widehat{=}\, A \rightarrowtail B \cap A \twoheadrightarrow B$$

Examples 7.85 Which of the following are bijections?

1. $\{1 \mapsto red, 2 \mapsto green, 3 \mapsto yellow\}$ with source set $\{1,2,3\}$ and target set $\{red, green, yellow\}$

2. $\{1 \mapsto red, 2 \mapsto green, 3 \mapsto yellow, 4 \mapsto red\}$ with source set $\{1,2,3,4\}$ and target set $\{red, green, yellow\}$

3. $\{1 \mapsto red, 2 \mapsto green, 3 \mapsto yellow\}$ with source set $\{1,2,3,4\}$ and target set $\{red, green, yellow\}$

4. $\{1 \mapsto red, 2 \mapsto green, 3 \mapsto yellow\}$ with source set $\{1,2,3\}$ and target set $\{red, green, yellow, blue\}$

5. $\{1 \mapsto red, 2 \mapsto green, 3 \mapsto green\}$ with source set $\{1, 2, 3\}$ and target set $\{red, green\}$

6. $\{1 \mapsto red, 2 \mapsto green, 3 \mapsto green\}$ with source set $\{1, 2, 3\}$ and target set $\{red, green, yellow\}$

Solution 7.86

1. $\{1 \mapsto red, 2 \mapsto green, 3 \mapsto yellow\}$ with source set $\{1, 2, 3\}$ and target set $\{red, green, yellow\}$ is total, injective and surjective; hence it is a bijection.

2. $\{1 \mapsto red, 2 \mapsto green, 3 \mapsto yellow, 4 \mapsto red\}$ with source set $\{1, 2, 3, 4\}$ and target set $\{red, green, yellow\}$ is total and surjective, but is not injective (both 1 and 4 map to *red*); hence it is not a bijection.

3. $\{1 \mapsto red, 2 \mapsto green, 3 \mapsto yellow\}$ with source set $\{1, 2, 3, 4\}$ and target set $\{red, green, yellow\}$ is both injective and surjective, but not total (4 is not in the domain); hence it is not a bijection.

4. $\{1 \mapsto red, 2 \mapsto green, 3 \mapsto yellow\}$ with source set $\{1, 2, 3\}$ and target set $\{red, green, yellow, blue\}$ is total and injective, but not surjective (*blue* is not in the range); hence it is not a bijection.

5. $\{1 \mapsto red, 2 \mapsto green, 3 \mapsto green\}$ with source set $\{1, 2, 3\}$ and target set $\{red, green\}$ is not injective as both 2 and 3 map to *green*; hence it is not a bijection.

6. $\{1 \mapsto red, 2 \mapsto green, 3 \mapsto green\}$ with source set $\{1, 2, 3\}$ and target set $\{red, green, yellow\}$ is neither injective (as previously shown) nor surjective (as *yellow* is not in the range).

Regarding a bijective function as a kind of relation we can see that it is in fact a 1–1 correspondence; each element of the source set is uniquely linked with an element of the target set. Note that if a bijection (that is a 1–1 correspondence) exists between two sets, then the sets must necessarily have the same number of elements.

7.7 Case study: project teams

We saw in section 5.7 that we could store information about the allocation of staff to project teams in the form of a relation, *project_member* $:\in People \leftrightarrow All\_Projects$. The value of *project_member* is equal to

$\{(Elma, B), (Elma, C), (Rajesh, A), (Rajesh, B), (Rajesh, D),$
$(Mary, B), (Mary, D), (Carlos, A), (Carlos, C), (Carlos, D),$
$(Mike, A), (Mike, B), (Mike, D)\}$

We also found that we could extract the *Workforce* information from this relation using *Workforce* $= \text{dom} \, project\_member$ provided each employee was allocated to at least one project. A difficulty arises, however, if a person is not currently allocated to a project. One way round the problem might be to have a null project to which anyone is allocated whenever they are not allocated to a real project.

An alternative way to overcome this difficulty is to choose the target set to be $\mathbb{P}\,All\_Project$; each person is related to the set of project teams they belong to. This new relation, *project_set*, for the information given in section 2.4 is equal to:

$$\{(Elma, \{B, C\}), (Rajesh, \{A, B, D\}), (Mary, \{B, D\}),$$
$$(Carlos, \{A, C, D\}), (Mike, \{B, D\})\}$$

In fact this new relation is a *total* function on *Workforce*, so that storing this function we shall not lose any information about the set *Workforce*.

If now we have a new employee, *Helga* say, who is not initially allocated to any project, this can be taken care of by mapping *Helga* to the empty set. The value of the function *project_set* now becomes:

$$\{(Elma, \{B, C\}), (Rajesh, \{A, B, D\}), (Mary, \{B, D\}),$$
$$(Carlos, \{A, C, D\}), (Mike, \{B, D\}), (Helga, \varnothing)\}$$

Now *project_set*$^\sim$ is not a function for either of the two values given here. Since both *Mary* and *Mike* map to $\{B, D\}$, the function is not injective; from definition 7.77 we know that the function cannot therefore have an inverse function. Of course there may be particular values of *project_set* which are injective, but in general the function is not injective. In handling this function, it may well be important to realize this. A procedure may be written which inadvertently assumes that *project_set*$^\sim$ is a function; worse, if test cases were not carefully chosen then we might only use injective values of *project_set* and a bug in the program might not be discovered. Clearly in designing the test data, we should need to include both injective and non-injective values for *project_set*.

Note how in these examples, we have been regarding functions, such as *project_set*, as entities which are variable; functions have a similar status to ordinary variables. This view is often expressed by describing functions as *first-class citizens*. Contrast this with the more elementary view in which a function is a fixed entity, even though the value obtained by *applying* the function is variable; in this more elementary view, the function is only a *second-class citizen*. The view of functions as first-class citizens lies very much at the heart of so-called *functional languages* such as *ML* and *Miranda*.

7.8   Exercise

1. Which of the following sets are functions?

    (a) $\{(1, 2), (2, 3), (3, 7), (7, 1)\}$

    (b) $\{(1, 2), (2, 3), (3, 7), (7, 1), (1, 5)\}$

(c) $\{(1,2)\}$

(d) $\{(1,2),(2,2),(3,2),(7,2)\}$

(e) $\{(1,2),(2,2),(3,2),(7,2),(1,2)\}$

(f) $\{((1,2),(2,3)),((3,7),(7,1))\}$

(g) $\{((1,2),(2,3)),((3,7),(7,1)),((1,5),(3,7))\}$

(h) $\{(\{2,3\},\{1\}),(\{1,2\},\{3\}),(\{1,3\},\{2\})\}$

(i) $\{\{(2,3),(1,4)\},\{(1,2),(3,4)\},\{(1,3),(2,4)\}\}$

(j) $\{(\{\},\{\})\}$

(k) $\{(Elma,A),(Elma,B),(Mike,C),(Rajesh,D)\}$

(l) $\{(Elma,A),(Carlos,B),(Mike,C),(Rajesh,D)\}$

(m) $\{(Big,Elma,A),(Big,Rajesh,B),(Little,Elma,C)\}$

(n) $\{(Big,(Elma,A)),(Big,(Rajesh,B)),(Little,(Elma,C))\}$

(o) $\{((Big,Elma),A),((Big,Rajesh),B),((Little,Elma),C)\}$

2. Find the domain and range of each of the following functions.

   (a) $\{2 \mapsto 4, 0 \mapsto 3, 1 \mapsto 7, 7 \mapsto 5\}$

   (b) $\{2 \mapsto 4, 0 \mapsto 3, 1 \mapsto 7, 3 \mapsto 0\}$

   (c) $\{1 \mapsto 3, 2 \mapsto 3, 3 \mapsto 3, 7 \mapsto 3\}$

   (d) $\{\}$

   (e) $\{(0 \mapsto 2, 6 \mapsto 4),(3 \mapsto 5, 7 \mapsto 1)\}$

   (f) $\{\{2,3\} \mapsto \{1\}, \{1,2\} \mapsto \{3\}, \{1,3\} \mapsto \{2\}\}$

   (g) $\{\{\} \mapsto \{\}\}$

   (h) $\{(Elma,A),(Mike,C),(Rajesh,D)\}$

   (i) $\{(Elma,A),(Carlos,B),(Mike,C),(Rajesh,D)\}$

   (j) $\{(Elma,A),(Carlos,B),(Mike,C),(Mary,D)\}$

   (k) $\{(Elma,A),(Carlos,B),(Mike,C),(Mary,D),(Rajesh,D)\}$

   (l) $\{(Elma,Big),(Carlos,Big),(Mike,Big)\}$

   (m) $\{((Big,Elma),A),((Big,Rajesh),B),((Little,Elma),C)\}$

3. Decide whether each of the following functions is total, injective, surjective, bijective, or a combination of these.

   (a) $\{0 \mapsto 3, 1 \mapsto 4, 2 \mapsto 5, 3 \mapsto 6\}$ with source set $\{0,1,2,3\}$, target set $\{3,4,5,6,7\}$

   (b) $\{3 \mapsto 0, 4 \mapsto 1, 5 \mapsto 2, 6 \mapsto 3\}$ with source set $\{3,4,5,6,7\}$, target set $\{0,1,2,3\}$

(c) $\{0 \mapsto 3, 1 \mapsto 4, 2 \mapsto 5, 3 \mapsto 6\}$ with source set $\{0, 1, 2, 3\}$, target set $\{3, 4, 5, 6\}$

(d) $\{0 \mapsto 3, 1 \mapsto 3, 2 \mapsto 5, 3 \mapsto 6\}$ with source set $\{0, 1, 2, 3\}$, target set $\{3, 4, 5, 6\}$

(e) $\{1 \mapsto 4, 2 \mapsto 5, 3 \mapsto 6\}$ with source set $\{0, 1, 2, 3\}$, target set $\{3, 4, 5, 6, 7\}$

(f) $\{0 \mapsto 3, 1 \mapsto 3\}$ with source set $\{0, 1, 2, 3\}$, target set $\{3, 4, 5, 6, 7\}$

(g) $\{\}$ with source set $\{2, 4, 6\}$, target set $\{3, 5, 7, 8\}$

(h) $\{\}$ with source set $\{2, 4, 6\}$, target set $\{\}$

(i) $\{\}$ with source set $\{\}$, target set $\{3, 5, 7, 8\}$

(j) $\{\}$ with source set $\{\}$, target set $\{\}$

(k) $\{(Elma, A), (Carlos, B), (Mike, C)\}$
with source set $\{Elma, Rajesh, Mary, Carlos, Mike\}$
and target set $\{A, B, C, D\}$.

(l) $\{(Elma, A), (Carlos, B), (Mike, C), (Mary, D)\}$
with source set $\{Elma, Rajesh, Mary, Carlos, Mike\}$
and target set $\{A, B, C, D\}$.

(m) $\{(Elma, A), (Carlos, B), (Mike, C), (Mary, D), (Rajesh, D)\}$
with source set $\{Elma, Rajesh, Mary, Carlos, Mike\}$
and target set $\{A, B, C, D\}$.

(n) $\{(Elma, Big), (Carlos, Big), (Mike, Big)\}$
with source set $\{Elma, Rajesh, Mary, Carlos, Mike\}$
and target set $\{Big, Little\}$.

(o) $\{(Elma, Big), (Carlos, Big), (Mike, Little)\}$
with source set $\{Elma, Rajesh, Mary, Carlos, Mike\}$
and target set $\{Big, Little\}$.

4. Evaluate each of the following; which are functions?

(a) $\{(2, 4), (0, 3), (1, 7), (7, 5)\}^{\sim}$

(b) $\{(2, 4), (0, 3), (1, 7), (3, 0)\}^{\sim}$

(c) $\{(1, 3), (2, 3), (3, 3), (7, 3)\}^{\sim}$

(d) $\{\}^{\sim}$

(e) $\{((0, 2), (6, 4)), ((3, 5), (7, 1))\}^{\sim}$

(f) $\{(\{2, 3\}, \{1\}), (\{1, 2\}, \{3\}), (\{1, 3\}, \{2\})\}^{\sim}$

(g) $\{(\{\}, \{\})\}^{\sim}$

(h) $\{(Elma, A), (Carlos, B), (Mike, C)\}^{\sim}$

(i) $\{(Elma, A), (Carlos, B), (Mike, C), (Mary, D)\}^{\sim}$

(j) $\{(Elma, A), (Carlos, B), (Mike, C), (Mary, D), (Rajesh, D)\}^{\sim}$

(k) $\{(Elma, Big), (Carlos, Big), (Mike, Big)\}^{\sim}$

(l) $\{(Elma, Big), (Carlos, Big), (Mike, Little)\}^{\sim}$

5. In the following, each function has source and target set equal to $\mathbb{Z}$. Evaluate

(a) $\{1 \mapsto 2, 2 \mapsto 3, 3 \mapsto 4, 4 \mapsto 5\} \circ \{4 \mapsto 1, 0 \mapsto 3, 2 \mapsto 3, 3 \mapsto 2, 5 \mapsto 4\}$

(b) $\{1 \mapsto 2, 2 \mapsto 3, 3 \mapsto 4, 4 \mapsto 3\} \circ \{4 \mapsto 1, 0 \mapsto 3, 2 \mapsto 3, 3 \mapsto 2, 5 \mapsto 4\}$

(c) $\{2 \mapsto 3, 3 \mapsto 4, 4 \mapsto 5\} \circ \{4 \mapsto 1, 0 \mapsto 3, 2 \mapsto 3, 3 \mapsto 2, 5 \mapsto 4\}$

(d) $\{2 \mapsto 3, 3 \mapsto 4, 4 \mapsto 5, 5 \mapsto 8\} \circ \{4 \mapsto 1, 0 \mapsto 3, 2 \mapsto 3, 3 \mapsto 2, 5 \mapsto 4\}$

(e) $\{2 \mapsto 4, 0 \mapsto 3, 1 \mapsto 7, 7 \mapsto 5\} \circ \{2 \mapsto 4, 0 \mapsto 3, 1 \mapsto 7, 7 \mapsto 5\}$

(f) $\{2 \mapsto 4, 0 \mapsto 3, 1 \mapsto 7, 7 \mapsto 5\} \circ \{\}$

(g) $\{\} \circ \{2 \mapsto 4, 0 \mapsto 3, 1 \mapsto 7, 7 \mapsto 5\}$

(h) $\{2 \mapsto 4, 0 \mapsto 3, 1 \mapsto 7, 7 \mapsto 5\}^{-1} \circ \{2 \mapsto 4, 0 \mapsto 3, 1 \mapsto 7, 7 \mapsto 5\}$

(i) $\{2 \mapsto 4, 0 \mapsto 3, 1 \mapsto 7, 3 \mapsto 0\}^{-1} \circ \{2 \mapsto 4, 0 \mapsto 3, 1 \mapsto 7, 3 \mapsto 0\}$

(j) $\{2 \mapsto 4, 0 \mapsto 3, 1 \mapsto 7, 7 \mapsto 5\} \circ \{2 \mapsto 4, 0 \mapsto 3, 1 \mapsto 7, 7 \mapsto 5\}^{-1}$

(k) $\{2 \mapsto 4, 0 \mapsto 3, 1 \mapsto 7, 3 \mapsto 0\} \circ \{2 \mapsto 4, 0 \mapsto 3, 1 \mapsto 7, 3 \mapsto 0\}^{-1}$

(l) $(\{4 \mapsto 1, 3 \mapsto 7, 7 \mapsto 5\} \circ \{2 \mapsto 4, 0 \mapsto 3, 1 \mapsto 7, 7 \mapsto 5\})^{-1}$

(m) $\{4 \mapsto 1, 3 \mapsto 7, 7 \mapsto 5\}^{-1} \circ \{2 \mapsto 4, 0 \mapsto 3, 1 \mapsto 7, 7 \mapsto 5\}^{-1}$

(n) $\{2 \mapsto 4, 0 \mapsto 3, 1 \mapsto 7, 7 \mapsto 5\}^{-1} \circ \{4 \mapsto 1, 3 \mapsto 7, 7 \mapsto 5\}^{-1}$

6. In the following, each function has source and target set equal to $\mathbb{Z}$. Evaluate

(a) $\{1 \mapsto 2, 2 \mapsto 3, 3 \mapsto 4, 4 \mapsto 5\} \oplus \{4 \mapsto 1, 0 \mapsto 3, 2 \mapsto 3, 3 \mapsto 2, 5 \mapsto 4\}$

(b) $\{4 \mapsto 1, 0 \mapsto 3, 2 \mapsto 3, 3 \mapsto 2, 5 \mapsto 4\} \oplus \{1 \mapsto 2, 2 \mapsto 3, 3 \mapsto 4, 4 \mapsto 5\}$

(c) $\{4 \mapsto 1, 0 \mapsto 3, 2 \mapsto 3, 3 \mapsto 2, 5 \mapsto 4\} \oplus \{4 \mapsto 1, 0 \mapsto 3, 2 \mapsto 3, 3 \mapsto 2, 5 \mapsto 4\}$

(d) $\{4 \mapsto 1, 0 \mapsto 3, 2 \mapsto 3, 3 \mapsto 2, 5 \mapsto 4\} \oplus \{2 \mapsto 3, 3 \mapsto 2, 5 \mapsto 4\}$

(e) $\{2 \mapsto 3, 3 \mapsto 2, 5 \mapsto 4\} \oplus \{4 \mapsto 1, 0 \mapsto 3, 2 \mapsto 3, 3 \mapsto 2, 5 \mapsto 4\}$

(f) $\{\} \oplus \{4 \mapsto 1, 0 \mapsto 3, 2 \mapsto 3, 3 \mapsto 2, 5 \mapsto 4\}$

(g) $\{4 \mapsto 1, 0 \mapsto 3, 2 \mapsto 3, 3 \mapsto 2, 5 \mapsto 4\} \oplus \{\}$

(h) $\{2 \mapsto 4, 0 \mapsto 3, 1 \mapsto 7, 7 \mapsto 5\}^{-1} \oplus \{2 \mapsto 4, 0 \mapsto 3, 1 \mapsto 7, 7 \mapsto 5\}$

(i) $(\{4 \mapsto 1, 3 \mapsto 7, 7 \mapsto 5\} \oplus \{2 \mapsto 4, 0 \mapsto 3, 1 \mapsto 7, 7 \mapsto 5\})^{\sim}$

(j) $\{4 \mapsto 1, 3 \mapsto 7, 7 \mapsto 5\}^{-1} \oplus \{2 \mapsto 4, 0 \mapsto 3, 1 \mapsto 7, 7 \mapsto 5\}^{-1}$

(k) $\{2 \mapsto 4, 0 \mapsto 3, 1 \mapsto 7, 7 \mapsto 5\}^{-1} \oplus \{4 \mapsto 1, 3 \mapsto 7, 7 \mapsto 5\}^{-1}$

7. Evaluate each of the following:

(a) $\{(Elma, A), (Carlos, B), (Mike, C)\}$
    $\oplus \{(Elma, A), (Carlos, B), (Mike, C), (Mary, D)\}$

(b) $\{(Elma, A), (Carlos, B), (Mike, C), (Mary, D), (Rajesh, D)\}^{\sim}$
    $\overset{\circ}{\underset{9}{}} \{(Elma, Big), (Carlos, Big), (Mike, Big)\}$

(c) $\{(Elma, Big), (Carlos, Big), (Mike, Little)\}$
    $\circ \{(Elma, A), (Carlos, B), (Mike, C)\}^{\sim}$

CHAPTER 8

Mathematical models

8.1 Introduction

There is little point in learning set theory and logic unless it has some use! Although examples of application are given throughout the book, it is worth exploring a little more the ways in which discrete mathematics is useful. Two aspects of discrete mathematics are particularly important: the use of mathematical models, and reasoning about the properties of models. To a certain extent you have been introduced to both modelling and reasoning throughout the book; indeed, it is intended that this *Essence* book will give you the necessary prerequisite knowledge, skills and attitudes to enable a proper study of these areas.

This chapter will look more closely at mathematical modelling, while a very brief introduction to reasoning is given in the concluding chapter. Hopefully you will be motivated to study further.

8.2 Mathematical models

We have already met the concept of mathematical models of functions and relations. In fact models are commonly used to capture the properties of systems; that is a mathematical model has properties analogous to those of the system it is modelling.

As a simple example, we shall consider traffic signals. In Great Britain, for example, traffic signals have the following cyclical sequence of *states*:

- green
- orange
- red
- red–orange
- green

We can begin building a model of this system by introducing the set *signal_colours* to represent the possible colours used in the signalling system. The first important property that we capture is the fact that there are precisely three distinct colours used.

$$\#signal\_colours = 3$$

This is a very simple model and has only captured one property. In order to explain other properties, we need to introduce *identifiers* for the three different colours. For example we could write

$$signal\_colours = \{x, y, z\}$$

Note that we must distinguish between the identifiers, that is the labels associated with the colours, and the colours themselves. It is necessary to make explicit the association between each identifier and the colour to which it refers. Thus we might say that the identifier x corresponds to the colour green, y to orange, and z to red. To make life easier, instead of x, y, z it would of course be better to use labels that help us to remember the colours to which they refer. In Great Britain, the labels normally used are '*green*', '*amber*' and '*red*'. It is now not really necessary to state that *green* corresponds to the colour green, *amber* to orange (although this does seem a little obscure!) and *red* to red; the correspondence is (more or less) obvious. Thus we write

$$signal\_colours = \{green, amber, red\}$$

Note that from a purely mathematical point of view, this statement does not tell us how many different colours there are; it could be the case, say, that the labels '*amber*' and '*red*' both referred to the same colour. For this reason we still need to stipulate that the cardinality of *signal_colours* is three; since there are only three labels used in the set enumeration, then we can deduce that each label does indeed refer to a separate colour.

Having now introduce some appropriate labels for the constants of the system, we can proceed to state other properties we want the system to have. Thus we might want to stipulate that at any time either one or two colours are lit. To do this, we introduce the variable *colours_lit* to represent the colours showing at any particular time. This is most easily done by defining *colours_lit* to be a subset of *signal_colours*. Although we could introduce a new symbol, such as $:\subseteq$, to declare that one set is a subset of another, we can achieve the same effect by using $:\in$ with the power set operator:

$$colours\_lit :\in \mathbb{P} \, signal\_colours$$

We can now introduce predicates to capture the required properties of our system.

For example, we know that at any time the number of lights showing is either one or two. Hence we can write:

$$\#colours\_lit = 1 \lor \#colours\_lit = 2$$

This model is said to be a *refinement* of the earlier model, because we have specified an additional property. Although this new property excludes some unwanted combinations of colours, other unwanted combinations are still possible. For instance, it is still possible for *colours_lit* to take the value $\{red, green\}$, that is for both red and green to show together. To eliminate this possibility, we can refine the model further by including yet another property:

$$colours\_lit \neq \{red, green\}$$

Examples 8.1 This model still needs one further property to avoid unwanted colour combinations. Write an appropriate statement.

Solution 8.2 The combination $\{amber, green\}$ is also not permissible:

$$colours\_lit \neq \{amber, green\}$$

Combining these with logical connectives gives the following predicate:

$$\#colours\_lit = 1 \lor \#colours\_lit = 2 \land$$
$$colours\_lit \neq \{red, green\} \land$$
$$colours\_lit \neq \{amber, green\}$$

This is a little complicated, though not nearly so cumbersome as writing down all the possible combinations of colours: $colours\_lit = \{green\} \lor colours\_lit = \{amber\} \lor colours\_lit = \{red\} \lor colours\_lit = \{red, amber\}$.

Examples 8.3 Write a simpler predicate to completely specify all possible colour combinations.

Solution 8.4 One possible solution is:

$$\#colours\_lit = 1 \lor colours\_lit = \{red, amber\}$$

which says that all colours may be lit individually, but the only combination allowed is red with amber.

8.3   Modelling abstract data types

An *abstract data type* can be thought of as the collection of properties that a required data structure and its associated operations must have but the details of how the data structure is to be implemented are not given. An abstract data type is often associated with some real world concept.

Software engineers with a sound theoretical knowledge and a good understanding of mathematics can happily think about abstract data types using what are known as *axioms*. Many people, however, find this purely abstract approach a little too hard to cope with, and would rather think in terms of an actual *data structure*, that is an implementation of an abstract data type. Unfortunately, a data structure using conventional computer programming constructs (indexed lists for example) have too much detail that is not really part of the essential properties of the abstract data type. A very good compromise is to build a 'structure' using sets; this structure is a set theoretic model of the abstract data type. An ideal model will enable all relevant information to be captured, without including any unwanted or irrelevant information.

One example is the concept of *multiset* or *bag*. A multiset or bag is a collection of objects, some of which may be repeated.

Thus we may have a bag containing the following:

- 15mm bolt
- 17mm bolt
- 18mm bolt
- 15mm bolt
- 17mm bolt

Special brackets $[\![$ and $]\!]$ are used to delimit a bag enumeration. Thus, for example we could write the example given as

$$[\![15, 17, 18, 15, 17]\!]$$

Note that repetitions in the bag enumeration *are* significant, although the order of items in the enumeration is *not*. Thus we have $[\![15, 17, 18, 15, 17]\!] = [\![15, 15, 17, 17, 18]\!]$ but $[\![15, 17, 18, 15, 17]\!] \neq [\![15, 17, 18]\!]$.

It would not be possible to model this collection of objects as the set $\{15, 17, 18, 15, 17\}$ since this is the same set as $\{15, 17, 18\}$. Hence not all relevant information would be captured.

An alternative model would be to use a function which maps each integer in the range $1 \ldots number\_of\_objects$ to an object. For example we *could* model $[\![15, 17, 18, 15, 17]\!]$ by $\{1 \mapsto 15, 2 \mapsto 17, 3 \mapsto 18, 4 \mapsto 15, 5 \mapsto 17\}$. This new model certainly captures all the relevant information about the bag, but unfortunately carries extra unwanted information, namely the order of the objects. In particular the sets $\{1 \mapsto 15, 2 \mapsto 17, 3 \mapsto 18, 4 \mapsto 15, 5 \mapsto 17\}$ and $\{1 \mapsto 15, 2 \mapsto 15, 3 \mapsto 17, 4 \mapsto 17, 5 \mapsto 18\}$ are not equal even though the bags to which they correspond *are* equal.

A better model is a function which maps each object in the bag to a positive integer, this integer representing the number of repetitions of the object. The example could be modelled by the following set of maplets:

$$\{15 \mapsto 2, 17 \mapsto 2, 18 \mapsto 1\}$$

As often happens in modelling, we often equate an object to the model of that object. In this case we write:

$$[\![15, 17, 18, 15, 17]\!] = [\![15, 15, 17, 17, 18]\!] = \{15 \mapsto 2, 17 \mapsto 2, 18 \mapsto 1\}$$

In fact, since we are really using a function to model a bag (or more precisely we are using the set model of that function), then we need to specify a set from which the objects in the bag have been chosen. Suppose, for example, that we have a set consisting of the various identifiers for motor parts; a bag taken from this set might then be used to represent the stock of such parts held by a garage. Of course it may well be that the garage is out of stock of some items, which therefore would not occur in the bag; but their absence is important to the person in charge of stock control. Thus in our (very simple) example of bolt sizes, the full set of sizes may be, for example, $\{15, 16, 17, 18\}$. Then the bag $[\![15, 17, 18, 15, 17]\!]$, represented by the set $\{15 \mapsto 2, 17 \mapsto 2, 18 \mapsto 1\}$, holds the information that 16mm bolts are out of stock.

In general a bag of objects chosen from a set X can be modelled by a partial function from X to the set of positive integers, N_1. Thus in our example, we can define a set *Bolt_Sizes* equal to $\{15, 16, 17, 18\}$ and a function

$$bolt\_stock :\in Bolt\_Sizes \nrightarrow N_1$$

to represent the stock of bolts at any time. The value of this function will vary to reflect the changes in stock held; for example at one time *bolt_stock* might equal $[\![15, 17, 18, 15, 17]\!]$, but at a later time this could change to, say, $[\![15, 17, 18, 17]\!]$ if one of the 15mm bolts is taken out of stock..

Of course this is not the only way in which we can build a model of bags without introducing unwanted properties.

Examples 8.5 Suggest another way of modelling bags using functions.

Solution 8.6 Bags could be modelled by a total function from the set of all possible types of object to N. For example:

$$bolt\_stock :\in Bolt\_Sizes \rightarrow N$$

This alternative model explicitly represents items out of stock by associating them with a zero value.

In fact our choice of model is largely determined by the ease with which we can also model operations. We shall now consider how we can model operations on bags.

The model of a bag we have chosen can be regarded as a set, as a relation, as a function, or as a special type of object in its own right. We can therefore use all the usual operations applicable to sets, relations, and functions to bags; but not all such operations will give a bag as an answer. What is required however are models of the operations which will be required for handling bags.

For example, we may want to know the set of all objects represented at least once in a bag ('in stock' items). Using our model of bags, this can be obtained as

the domain of the function representing the bag. For example, the *in_stock* items
for the bag considered above are given by

$$in\_stock \, [\![15, 17, 18, 15, 17]\!]$$
$$= \mathrm{dom} \; [\![15, 17, 18, 15, 17]\!]$$
$$= \mathrm{dom} \, \{15 \mapsto 2, 17 \mapsto 2, 18 \mapsto 1\}$$
$$= \{15, 17, 18\}$$

The set of 'out of stock' items can be obtained by using set difference:

$$out\_of\_stock \, [\![15, 17, 18, 15, 17]\!]$$
$$= Bolt\_Sizes \setminus \mathrm{dom} \; [\![15, 17, 18, 15, 17]\!]$$
$$= Bolt\_Sizes \setminus \mathrm{dom} \, \{15 \mapsto 2, 17 \mapsto 2, 18 \mapsto 1\}$$
$$= Bolt\_Sizes \setminus \{15, 17, 18\}$$
$$= \{15, 16, 17, 18\} \setminus \{15, 17, 18\}$$
$$= \{16\}$$

Notice that if we had chosen to model a bag as a total function mapping to $\mathbb{N}$ rather
than a partial function mapping to $\mathbb{N}_1$, these operations could not be modelled so
easily. For example, taking the domain of the total function would not necessarily
give the 'in stock' items:

$$\mathrm{dom} \{15 \mapsto 2, 16 \mapsto 0, 17 \mapsto 2, 18 \mapsto 1\} = Bolt\_Sizes$$

Examples 8.7 The Kingdom of Erehwemos is a little known country whose local
currency is Erehwemon dollars. Suppose a particular design of vending machine,
destined for that country, takes coins with denominations of \$1, \$0.50, \$0.20, \$0.10
and \$0.05. We can model this by the set *Denominations* $= \{100, 50, 20, 10, 5\}$.

A collection of coins in a vending machine may then be modelled by a bag
drawn from the set *Denominations*. For example, shortly after being emptied
two vending machines may contain coins modelled by bags B_1 and B_2 where
$B_1 = [\![100, 20, 100, 50, 50, 50]\!]$ and $B_2 = [\![10, 20, 20]\!]$.

Find the *set* corresponding to each of the following, and state whether this set
could represent a bag taken from *Denominations*.

1. $\mathrm{dom} \, B_1$
2. $B_1 \cup B_2$
3. $B_1 \cap B_2$
4. $\mathbb{P} \, B_1$
5. $B_1 \oplus B_2$

Solution 8.8

1. $\mathrm{dom} \, B_1 = \{100, 50, 20\}$. This is the set of coin types held by the first
   machine. It is not a bag in itself.

2. $B_1 \cup B_2 = \{100 \mapsto 2, 50 \mapsto 3, 20 \mapsto 1, 20 \mapsto 2, 10 \mapsto 1\}$. This does not represent a bag since it is not even a function (20 maps to both 1 and 2).

3. $B_1 \cap B_2 = \varnothing = [\![\,]\!]$. Note that the empty set models the empty bag; in this case it represents a collection of no coins, such as might be found after emptying a vending machine. Indeed the empty set is frequently used to model the initial or reset value of a system. Note that in general it is always true that the intersection of two bags will be a bag.

4. $\mathbb{P} B_1 = \{\{20 \mapsto 2, 10 \mapsto 1\}, \{20 \mapsto 2\}, \{10 \mapsto 1\}, \{\}\}$. This is in fact a set of bags, namely $\{[\![10, 20, 20]\!], [\![20, 20]\!], [\![10]\!], [\![\,]\!]\}$.

5. $B_1 \oplus B_2 = \{100 \mapsto 2, 50 \mapsto 3, 20 \mapsto 2, 10 \mapsto 1\}$
   $= [\![100, 100, 50, 50, 50, 20, 20, 10]\!]$.

8.4   Modelling changes of state

In the traffic signal example, we were only able to model the possible *states* of the signal. The state of the signal at any instant is modelled by the set *colours_lit*. Usually we would also be interested to know what the possible transitions were between states. For example, with just the green lit, the only possible change in state would be from green to amber.

We can model valid changes of state by a relation; an element of the relation will be an ordered pair of the form (*state, state'*) where *state* represents the state before the transition and *state'* that after the transition.

Thus in the traffic signal example, we could model the allowable changes of state by a relation $\Delta State$ on the set *signal_colours*; the value of this relation is

$$\{(\{green\}, \{amber\}), (\{amber\}, \{red\}),$$
$$(\{red\}, \{red, amber\}), (\{red, amber\}, \{green\})\}$$

8.5   Exercise

A common type of data structure is a *list*. Indeed we have already met lists in set enumeration. The essential difference between a list and a set is that the order of elements in a list *is* important; and so are repetitions. Thus the lists

Carlos, Hannah, Mike
Mike, Carlos, Hannah
Carlos, Hannah, Mike, Hannah

are all different, even though the corresponding set enumerations

$\{Carlos, Hannah, Mike\}$
$\{Mike, Carlos, Hannah\}$
$\{Carlos, Hannah, Mike, Hannah\}$

are all equal. To emphasize that a list is indeed a list, we frequently use single square brackets to enclose the list: $[\ldots]$. For example we can write

$[Carlos, Hannah, Mike, Hannah]$

We frequently wish to attach one list to the end of another list; this operation is called *concatenation*. For example, we might list participants at a meeting in the order in which they booked to attend. Now suppose that prior to the meeting we have:

$Prior\_Bookings = [Nasrin, Adelaide, Jiri]$

and that on the day of the meeting we have:

$Late\_Bookings = [Maria, Robert]$

then the list of *Participants* at the meeting is given by the concatenation of *Late_Bookings* to the end of *Prior_Bookings*:

$Participants = [Nasrin, Adelaide, Jiri, Maria, Robert]$

A list can be modelled by a function $S : \mathbb{N} \nrightarrow A$ where A is a set from which the list objects may be drawn and $\operatorname{dom} S = 1 \ldots n$ where n is the length of the list. Such a function is called a *sequence* on A. For example if $A = \{a, b, c\}$ then the list a, b, c, a, a, b can be modelled by the set

$$\{(1, a), (2, b), (3, c), (4, a), (5, a), (6, b)\}$$

which is a sequence on $\{a, b, c\}$. The first member of each ordered pair gives the position in the list (or sequence) of the second member of the ordered pair. A convenient notation for sequences is to use angle brackets. Thus we could write

$$\langle a, b, c, a, a, b \rangle = \{(1, a), (2, b), (3, c), (4, a), (5, a), (6, b)\}$$

to represent the sequence. In the following questions take the values of two sequences S_1 and S_2 to be $S_1 = \langle a, b, a, a, c \rangle$ and $S_2 = \langle c, b, c \rangle$.

1. What are

   (a) $S_1(3)$

   (b) $\operatorname{dom} S_1$

   (c) $\operatorname{ran} S_2$

2. Find the set enumerations corresponding to each of the following; where the result is itself a sequence give the answer in the form $\langle \ldots \rangle$. Suggest appropriate applications for the various list operations modelled.

    (a) S_1

    (b) $S_1{}^{\sim}$

    (c) $S_1 \oplus S_2$

    (d) $S_2 \oplus S_1$

    (e) $S_1 \cup S_2$

    (f) $S_1 \cap S_2$

    (g) $S_1 \, {}_9^8 \, S_2{}^{\sim}$

    (h) $S_1 \, {}_9^8 \, S_2{}^{\sim} \, {}_9^8 \, S_1$

    (i) $S_2 \, {}_9^8 \, S_2{}^{\sim}$

    (j) $S_2 \, {}_9^8 \, S_2{}^{\sim} \, {}_9^8 \, S_2$

3. Write out the set enumeration for $\mathbb{P} S_2$. State how many elements of $\mathbb{P} S_2$ are sequences, and write these sequences in the form $\langle \ldots \rangle$.

4. The concatenation of two lists is to be modelled by a binary operator $^\frown$ on sequences. Obtain a suitable definition for $^\frown$ in terms of operations on sets, relations, and functions.

CHAPTER 9

Quo vadis?

9.1 Review

In this book only the *essence* of discrete mathematics has been considered, namely sets and logic (propositional and predicate), and their simple application to relations and functions. Topics such as 'Graph Theory' or 'Combinatorics' have been purposely left out in the belief that once a sound understanding of sets and logic has been achieved, then these other areas can be very easily learnt later. Formal proof has been alluded to but not fully developed; it seems unlikely that many readers will ever need to become involved with formal proof. But modelling is something that everyone does, perhaps without realizing it. What I have tried to show in this book is the value of modelling with sets and logic, and to build the skills and understanding that are a necessary pre-requisite for such modelling. I believe that such a modelling ethos is the principal reason for studying discrete mathematics.

The approach adopted has been to develop mental models of the mathematical concepts in terms of programming, or *algorithmic*, concepts without becoming too concerned with their application in the early stages. Later the application through modelling has been covered briefly. Becoming a proficient modeller takes time and effort, and it would be unrealistic to expect the reader to develop such skills from one short book; the reference to application has therefore been included to justify the need for discrete mathematics, and to give a brief introduction.

At this stage it is worth reviewing the algorithmic approach to sets and then the nature of modelling.

9.1.1 Sets as algorithms

The basic concept that we have used is that of set membership; we have supposed that a set can be associated with a process which determines whether any given object is a member of the set or not. In mathematical terms the algorithm can be given either as a list (set enumeration) or using the notation of predicate logic (set comprehension). It should be stressed that this view of sets is not without fault; in more advanced texts, the reader will learn that not all sets can be associated with *decision procedures*. In spite of its shortcomings, however, the approach presented in this book does help to establish some basic understanding and familiarity with set theory upon which more advanced mathematical ideas can be built.

From the basic concept of set membership all the other concepts in this book have been developed: set operations and relations, such as the union and intersection operations and the subset relation; subsequently these have been used to model functions and relations, which in turn can be used to model systems and data types. Thus we can regard logic as the assembly level language of a 'set oriented' processing unit. In fact set based languages do exist, of which ISETL is one of the more commonly available. ISETL is short for 'Interactive Set Language'. This language enables the programmer to program in terms of sets and set operations. In addition, other languages such as the functional languages (such as Miranda) and logic languages (such as Prolog) depend very much upon the ideas of discrete mathematics.

9.1.2   Modelling

Throughout the book, examples have been given of how the ideas of discrete mathematics can be applied and a whole chapter has been devoted to some simple modelling. At first the idea of modelling a real system with sets and set operations may seem a little weird. However, it is no more difficult than reading a map. Most people have learnt to read a map at some stage in their lives, and eventually do not worry too much about the fact that they are dealing with what is really a mathematical model, complete with funny symbols and conventions! Of course some people seem to be better at map reading than others, yet no one would deny that maps *are* very useful, and that most people can use them to a greater or lesser extent. It should also be borne in mind that drawing maps is a rather more difficult skill than reading them, but most people can at least draw very simple maps – for example to let a friend know the way to their house. Modelling with discrete mathematics is essentially no different; if you have successfully worked your way through this book, then you should at least be a modestly competent 'map-reader' and recognize the value of your 'map-reading' skills.

9.2   Further study

Hopefully, the reader will want to study further, and to develop both the mathematical and modelling skills the foundations of which should have been laid by working through this book. This section is intended to be a guide to some of these areas and contains very brief outlines of what they are concerned with. Further information can be found in numerous books or from the 'internet' in newsgroups and virtual libraries.

9.2.1   Denotational semantics

In the early days of electronic computers, programming languages tended to be machine specific. The action of any given command or construct in the language could be easily defined in terms of how it caused the relevant machine to behave. With the advent of general languages, usable on a variety of platforms, it became necessary to devise an unambiguous way of defining the action of each command in a high level language. Several approaches have been tried, but the one that has proved most successful has been *denotational semantics*. This approach is based upon set theory, though is somewhat more sophisticated than the simple modelling of data types and operations discussed in this book. Nevertheless, you should by now realize the value of discrete mathematics in defining operations unambiguously.

9.2.2   Type theory

The theory of types had its origins in the work of Bertrand Russell at the beginning of the twentieth century; it has developed into an extensive theory which, as the name suggests, does have some connection with data types in programming languages and with the typed predicate logic used in this book.

9.2.3   Formal methods

A mathematical model is intended to capture all the relevant properties of a system. In all but the simplest, however, it is impractical to list *all* the required properties explicitly. Furthermore, there is always the danger that some important property has been overlooked. We therefore need to be able to *reason* about the model.

Take for example the traffic lights system. We have seen how the possible values of *colours_lit* may be given explicitly as

$$colours\_lit = \{green\} \vee colours\_lit = \{amber\} \vee colours\_lit = \{red\}$$
$$\vee \; colours\_lit = \{red, amber\}$$

but that a more concise expression of this model is

$$\#colours\_lit = 1 \vee colours\_lit = \{red, amber\}$$

We need to convince ourselves that the second statement of the model is equivalent to the first. We can argue this by use of appropriate set notation and using properties of sets that, by now, should be familiar.

Introduce the set *Possible_States* to include all the valid sets of colours lit:

$$Possible\_States = \{\{green\}, \{amber\}, \{red\}, \{red, amber\}\}$$

Then we can argue as follows:

> *Possible_States*
> $= \{\{green\}, \{amber\}, \{red\}, \{red, amber\}\}$
> $= \{\{green\}, \{amber\}, \{red\}\} \cup \{\{red, amber\}\}$
> $= \{colours\_lit :\in \; \mathbb{P}\{green, amber, red\} \mid \#colours\_lit = 1\}$
> $\cup$
> $\{colours\_lit :\in \; \mathbb{P}\{green, amber, red\} \mid colours\_lit = \{red, amber\}\}$
> $= \{colours\_lit :\in \; \mathbb{P}\{green, amber, red\} \mid$
> $\qquad \#colours\_lit = 1 \vee colours\_lit = \{red, amber\}\}$

Reasoning using the established properties of sets is a very useful and powerful technique; the use of *formal logic* enables the use of software tools to help the process. It is, nevertheless, an extensive topic which requires another book in its own right, and cannot be dealt with adequately in this short book. Nevertheless the 'flavour' of formal logic can be given by means of another example.

Examples 9.1 Show that if X is any arbitrary set, then $X \times \varnothing = \varnothing$.

Solution 9.2 Using knowledge about the properties of sets and some elementary arithmetic we can write down some rules:

1. If $\#S = 0$ then we can write down $S = \varnothing$ for any set expression S; this expresses the uniqueness of the empty set.
2. $n \times 0 = 0$ for any number n.
3. $\#\varnothing = 0$
4. $\#(A \times B) = \#A * \#B$

We can now apply these rules:

$$
\begin{array}{rcll}
\#(X \times \varnothing) = & \#X * \#\varnothing & \text{Rule 4} \\
= & \#X * 0 & \text{Rule 3} \\
= & 0 & \text{Rule 2} \\
X \times \varnothing = & \varnothing & \text{Rule 1}
\end{array}
$$

In formal logic, deductions are made by applying strictly defined rules to strings of characters to generate further strings of characters; in this manner it *should* be possible to carry out complicated proofs and derivations of mathematical theorems. Russell and Whitehead did indeed try this in the early twentieth century, but quickly discovered that even a simple theorem like '$1 + 1 = 2$' required an enormous amount of work. Now a computer program can be regarded as a mathematical theorem which can, in principle, be proved from a specification, always provided that the specification is itself written in appropriate formal notation. For many years there have been efforts to develop programs from specifications using such *formal methods* – a term borrowed from logic. This approach has had varying degrees of success. Even where no formal derivation or proof of a program has been practical, the act of producing a formal specification is often found to enhance the reliability and reduce the overall cost of software production and maintenance. Such formal specifications are based on logic and often also on set theory. One particular notation, 'Z', is a good notation for expressing set theoretic models; it is very close to the notation used in this book. Other formal specification languages include VDM, B, Larch, LOTOS, CSP and CCS.

9.2.4   Mathematical theories

There are many branches of mathematics that are relevant to software and hardware development. A few of these are briefly described.

Graph theory
Graph theory applies to networks; not surprisingly it is important in computing and telecommunications. A graph consists of points or *nodes* which are linked to other nodes by *edges* or *arcs*. Any problem which can be thought of in terms of a network or a tree (a binary tree for example) can be modelled using graph theory, and the many powerful theorems of graph theory used.

Combinatorics
Combinatorics is concerned with the ways in which choices of combinations can be made. It is an area closely related to graph theory.

Coding and error correction
In communications, data is encoded as strings of *bits*. Sometimes the data becomes corrupted; in order to overcome such problems, techniques of error checking and error correcting codes have been developed. A notable example of the use of error correcting codes is in compact discs. Another related area is that of data encryption, necessary for the secure transmission and storage of sensitive information such as 'cash machines' at banks. All of these depend upon discrete mathematics.

Abstract algebra
Abstract algebra is concerned with defining the properties of operations; in tradi-tional 'school' algebra, these operations are those of ordinary arithmetic, but many other *algebras* exist. Each algebra consists of certain laws or *axioms* which describe the properties of one or more operations and constants of a system. For example the operation of concatenation, $^\frown$, on character strings satisfies certain properties of a kind of algebra known as a *monoid*: if x, y and z are any three character strings, and if $[]$ denotes the empty string, then the following three properties are true:

1. Associativity $(x \frown y) \frown z = x \frown (y \frown z)$
2. Left Identity $[] \frown x = x$

3. Right Identity $x \frown [] = x$

The specification of the properties required of a system, or an object class (in object oriented design), can be regarded as an algebra. Some formal specification languages such as Larch or LOTOS, for example, are based upon this approach and are often called algebraic languages. The operations of relational databases depend upon *relational algebras*.

Universal algebra and category theory
There is a branch of mathematics known as *universal algebra* which is concerned with the study of algebras themselves. It is finding increasing application in the computing world. For example building a more complex algebra from simpler ones finds applications in a modular approach to system specification and implementation, and in class inheritance in object oriented design.

Category theory is based upon the notion of *categories*; a category consists of *objects* and *arrows*, with each object being associated with a set of arrows. Categories are often used to describe objects and transformations on objects; an example is the category Set which has sets as the objects and functions between sets as the arrows. (There is much in common between universal algebra and category theory.)

Self-test questions

Each of the following questions has **one** correct answer. Choose what you consider to be the correct answer in each case.

Q.1 $\{0\} \cup \{1\}$ is equal to

    A $\varnothing$

    B $\{\varnothing\}$

    C $\{0, 1\}$

    D $\{\{0\}, \{1\}\}$

    E $\{\{0, 1\}\}$

Q.2 $\{\{0, 1, 2\}\} \cap \{\{0\}, \{1\}, \{2\}\}$ is equal to

    A $\varnothing$

    B $\{\varnothing\}$

    C $\{0, 1, 2\}$

    D $\{\{0\}, \{1\}, \{2\}\}$

    E $\{\{0, 1, 2\}, \{0\}, \{1\}, \{2\}\}$

Q.3 $\{3\} \times \{4, 5\}$ is equal to

    A 2

    B $\{12, 15\}$

    C $\{\{4, 3\}, \{5, 3\}\}$

    D $\{(3, 4), (3, 5)\}$

    E $\{(4, 3), (5, 3)\}$

Q.4 $\mathbb{P}\{\{5\}\}$ is equal to

    A $\{\{\{5\}\}\}$

    B $\{\varnothing, \{5\}\}$

    C $\{\{\varnothing, \{5\}\}\}$

    D $\{\{\varnothing\}, \{5\}\}$

    E $\{\varnothing, \{\{5\}\}\}$

Q.5 $\#(\bigcup\{\{0, 1, 2\}, \{0, 3\}, \{0, 2, 5\}\})$ is equal to

    A 0

    B 1

    C 3

    D 5

    E 7

Q.6 The truth table for $\neg p \Leftrightarrow (\neg q \vee p)$ is

A

p	q	$\neg p \Leftrightarrow (\neg q \vee p)$
T	T	T
T	F	F
F	T	F
F	F	F

B

p	q	$\neg p \Leftrightarrow (\neg q \vee p)$
T	T	F
T	F	F
F	T	F
F	F	T

C

p	q	$\neg p \Leftrightarrow (\neg q \vee p)$
T	T	T
T	F	F
F	T	F
F	F	T

D

p	q	$\neg p \Leftrightarrow (\neg q \vee p)$
T	T	T
T	F	T
F	T	T
F	F	T

E

p	q	$\neg p \Leftrightarrow (\neg q \vee p)$
T	T	T
T	F	T
F	T	F
F	F	T

Q.7 Which one of the following propositional forms is a contradiction?

    A $p \vee \neg p$

    B $p \Rightarrow \neg p$

    C $\neg p \Rightarrow \neg p$

    D $\neg p \Rightarrow p$

    E $p \Leftrightarrow \neg p$

Q.8 Which one of the following propositional forms is a tautology?

    A $p \wedge \neg q$

    B $p \vee \neg q$

    C $p \Rightarrow (p \wedge q)$

    D $p \Rightarrow (p \vee q)$

    E $\neg p \Rightarrow q$

Q.9 $p \Leftrightarrow q$ is equivalent to

    A $p \wedge q$

    B $\neg p \wedge \neg q$

    C $\neg p \Rightarrow \neg q$

    D $(p \Rightarrow q) \vee (q \Rightarrow p)$

    E $(p \vee q) \Rightarrow (p \wedge q)$

Q.10 $p \Rightarrow (q \Rightarrow r)$ is equivalent to

    A $(p \wedge q) \Rightarrow r$

    B $(p \vee q) \Rightarrow r$

    C $(p \Rightarrow q) \Rightarrow r$

    D $p \Rightarrow (q \wedge r)$

    E $p \Rightarrow (q \vee r)$

Q.11 The set enumeration for $\{x :\in \mathbb{Z} \mid x^2 \leqslant 9\}$ is

    A $\{1, 2, 3\}$

    B $\{0, 1, 2\}$

    C $\{0, 1, 2, 3\}$

    D $\{-2, -1, 0, 1, 2\}$

    E $\{-3, -2, -1, 0, 1, 2, 3\}$

Q.12 The set enumeration for $\{x :\in \mathbb{N}_1 \mid x < 5 \bullet (x - 1)\}$ is

    A $\{1, 2, 3\}$

    B $\{1, 2, 3, 4\}$

    C $\{0, 1, 2, 3\}$

    D $\{0, 1, 2, 3, 4\}$

    E $\{2, 3, 4, 5, 6\}$

Q.13 In which one of the following expressions is x a free variable?

    A $\{x :\in \mathbb{Z} \bullet x^2\}$

    B $\forall x :\in \mathbb{Z} \bullet x = 2$

    C $\forall y :\in \mathbb{Z} \bullet y = x$

    D $\exists x :\in \mathbb{Z} \bullet y = x$

    E $\exists x :\in \mathbb{Z} \bullet x = 2$

Q.14 Which one of the following propositions is true?

    A $\forall x :\in \mathbb{N} \mid x > 2 \bullet x < 4$

    B $\forall x :\in \mathbb{N} \mid x > 2 \bullet \neg(x < 4)$

    C $\forall x :\in \mathbb{N} \mid \neg(x > 2) \bullet x < 4$

    D $\forall x :\in \mathbb{N} \mid \neg(x > 2) \bullet \neg(x < 4)$

    E $\neg(\exists x :\in \mathbb{N} \mid x > 2 \bullet x < 4)$

Q.15 Which one of the following expressions is true for any choice of both the set A and predicate $P(x)$?

    A $\forall x :\in A \bullet (P(x) \Rightarrow \neg P(x))$

    B $(\forall x :\in A \bullet P(x)) \Rightarrow \neg(\forall x :\in A \bullet P(x))$

    C $(\forall x :\in A \bullet P(x)) \Rightarrow (\forall x :\in A \bullet \neg P(x))$

    D $(\forall x :\in A \bullet P(x)) \Rightarrow \neg(\forall x :\in A \bullet \neg P(x))$

    E $\neg(\forall x :\in A \bullet P(x)) \Rightarrow (\forall x :\in A \bullet \neg P(x))$

Q.16 Suppose we are given the relation $R :\in \mathbb{Z} \leftrightarrow A$ where the set $A = \{0, 1, 2\}$ and $R = \{(1, 1), (2, 1)\}$. Then the domain of this relation, $\operatorname{dom} R$, is equal to

    A $\varnothing$

    B $\mathbb{Z}$

    C $\{1\}$

    D $\{1, 2\}$

    E $\{0, 1, 2\}$

Q.17 Suppose we are given the following sets:

$$Colours = \{yellow, black, white\}$$
$$Primary = \{blue, green, red\}$$

Further suppose we are given the relation *Contains* between *Colours* and *Primary* such that *Contains* is equal to

$\{(white, green), (white, red), (white, blue), (yellow, red), (yellow, green)\}$

Then *Contains* $\S$ *Contains*$^\sim$ is equal to

A {*yellow, black*}

B {*yellow, black, white*}

C {(*yellow, yellow*) , (*white, white*)}

D {(*yellow, yellow*) , (*white, white*) , (*black, black*)}

E {(*yellow, yellow*) , (*white, white*) , (*yellow, white*) , (*white, yellow*)}

Q.18 Which one of the following sets could represent a transitive relation?

A $\varnothing$

B $\{\{1,2\},\{2,3\},\{1,3\}\}$

C $\{(1,2),(2,1)\}$

D $\{(1,2),(2,3),(3,1)\}$

E $\{(1,2),(2,3),(3,4),(1,4)\}$

Q.19 Suppose we are give a homogeneous relation S on positive integers in which

$$S = \{(1,2),(2,5),(4,6),(6,4)\}$$

The transitive closure S^+ is equal to

A $\{(1,5),(4,4)\}$

B $\{(1,5),(4,4),(6,6)\}$

C $\{(1,5),(2,2),(4,4),(6,6)\}$

D $\{(1,5),(2,2),(4,4),(5,1),(6,6)\}$

E $\{(1,2),(2,5),(1,5),(4,6),(6,4),(4,4),(6,6)\}$

Q.20 Which of the following statements is true for *any* homogeneous relation T?

A $\operatorname{dom} T = \operatorname{ran} T$

B $T \mathbin{\substack{\circ\\\circ}} T$ is transitive.

C $T \mathbin{\substack{\circ\\\circ}} T^{\sim}$ is transitive.

D $T \mathbin{\substack{\circ\\\circ}} T^{\sim}$ is reflexive.

E $T \mathbin{\substack{\circ\\\circ}} T^{\sim}$ is symmetric.

Q.21 Which one of the following expressions could represent a function?

A $(1,2)$

B $\{1,2\}$

C $\{\{1,2\},\{3,2\}\}$

D $\{(1,2),(3,2)\}$

E $\{(1,2),(1,3)\}$

Q.22 Suppose $f :\in \mathbb{N} \twoheadrightarrow \mathbb{P}\mathbb{N}$ is equal to $\{(4, \{2\}), (6, \{2, 3\}), (8, \{2\})\}$. Then $f(8)$ is equal to

 A 2

 B $\{2\}$

 C 8

 D $(8, 2)$

 E $(8, \{2\})$

Q.23 Which one of the following sets represents a surjection from $\{2, 3, 5, 7\}$ to $\{1, 2, 3\}$?

 A $\{\{2, 3, 5, 7\} \mapsto \{1, 2, 3\}\}$

 B $\{2 \mapsto 1, 5 \mapsto 3\}$

 C $\{1 \mapsto 2, 2 \mapsto 3, 3 \mapsto 7\}$

 D $\{2 \mapsto 1, 3 \mapsto 2, 7 \mapsto 3\}$

 E $\{2 \mapsto 1, 3 \mapsto 1, 5 \mapsto 2, 7 \mapsto 2\}$

Q.24 Suppose $f :\in \mathbb{N} \twoheadrightarrow \mathbb{N}$ is equal to $\{(1, 2), (2, 3), (3, 1), (4, 1)\}$ and $g :\in \mathbb{N} \twoheadrightarrow \mathbb{N}$ is equal to $\{(1, 1), (2, 4), (4, 3)\}$. Then $g \oplus f$ is equal to

 A $\{(3, 1)\}$

 B $\{(1, 1), (2, 4), (4, 3)\}$

 C $\{(1, 1), (2, 4), (3, 1), (4, 3)\}$

 D $\{(1, 2), (2, 3), (3, 1), (4, 1)\}$

 E $\{(1, 1), (1, 2), (2, 3), (2, 4), (3, 1), (4, 1), (4, 3)\}$

Q.25 Which one of the following sets will always represent a function whenever the set h represents a function? (Note that '$\times$' is used to denote the Cartesian product of sets.)

 A $h \times h$

 B $h^{\sim}$

 C $h \circ h^{\sim}$

 D $h^{\sim} \circ h$

 E $h^{\sim} \circ h \circ h^{\sim}$

Answers to exercises

B.1   Introduction to sets

1. (b), (f)
2. (a), (b), (c), (d), (e), (g), (h), (k), (l), (m)
3.   (a) 3

   (b) 3

   (c) 1

   (d) 3

   (e) 4
4.   (a) $\{\{9\}, \varnothing\}$

   (b) $\{\{\{9\}, \varnothing\}, \{\{9\}\}, \{\varnothing\}, \varnothing\}$

   (c) $\{\{3, 5\}, \{3\}, \{5\}, \varnothing\}$

   (d) $\{\{3, 5\}, \{3\}, \{5\}, \varnothing\}$
5.   (a) $\{3, 5, 9\}$

   (b) $\{1, 2, 3, 4, 5, 6, 9\}$

   (c) $\{2, 9\}$

   (d) $\{\{1\}, \{2\}, \{9\}\}$

   (e) $\{1, 9, \{2\}, \{9\}\}$

   (f) $\{1, 2, 3, 4, 5, 9\}$

   (g) $\{1, 2, 3, 4, 5, 9\}$

   The last two answers are equal. In general it is true that $X \cup (Y \cup Z) = (X \cup Y) \cup Z$; often we simply write $X \cup Y \cup Z$ since the order makes no difference to the answer.
6.   (a) $\{5\}$

   (b) $\varnothing$

   (c) $\{9, 2\}$

   (d) $\{\{9\}\}$

   (e) $\varnothing$

(f) $\{3\}$

(g) $\{3\}$

The last two answers are equal. In general it is true that $X \cap (Y \cap Z) = (X \cap Y) \cap Z$; often we simply write $X \cap Y \cap Z$ since the order makes no difference to the answer.

7. (a) $\{9\}$

(b) $\{9, 5, 2, 6\}$

(c) $\varnothing$

(d) $\{\{2\}\}$

(e) $\{9, \{2\}\}$

(f) $\{9\}$

(g) $\{9, 3\}$

The last two answers are different; the order in which the relative difference operator is applied influences the result. Note that the order in which subtraction is carried out in ordinary arithmetic also influences the result.

8. (a) $\{(9, 5), (9, 3), (5, 5), (5, 3)\}$

(b) $\{(9, 2), (9, 9), (2, 9), (2, 2)\}$

(c) $\{(\{9\}, \{9\}), (\{9\}, \{1\}), (\{2\}, \{9\}), (\{2\}, \{1\})\}$

(d) $\{(9, \{9\}), (9, 1), (\{2\}, \{9\}), (\{2\}, 1)\}$

(e) $\{((9, 3), 1), ((9, 3), 2), ((3, 3), 1), ((3, 3), 2)\}$

(f) $\{(9, (3, 1)), (9, (3, 2)), (3, (3, 1)), (3, (3, 2))\}$

The last two results are not equal.

9. (a) 4

(b) 3

(c) 1

(d) 1

(e) 4

10. If we take, for example, $X = \{1, 2\}$ and $Y = \{2, 3, 4\}$ then $\#X = 2$ and $\#Y = 3$.

(a) $\#(X \times Y) = 6 = 2 * 3 = \#X * \#Y$

(b) $\#\mathbb{P}X = 4 = 2^2 = 2^{\#X}$ and $\#\mathbb{P}Y = 8 = 2^3 = 2^{\#Y}$

(c) $\#X(\cap Y) = 1$ whereas $\#(X \cup Y) = 4 = 2 + 3 - 1 = \#X + \#Y - \#(X \cap Y)$

(d) $\#(X \setminus Y) = 1 = 2 - 1 = \#X - \#X(\cap Y)$

11. (a) $\{9\}$

(b) $\{9, 3\}$

(c) $\{9\}$

(d) $\{9,3\}$. Note that in general $(X \cup Y) \cap Z = (X \cap Z) \cup (Y \cap Z)$ while $(X \cap Y) \cup Z = (X \cup Z) \cap (Y \cup Z)$.

(e) $\{\{9,3,1\},\{3,1\},\{9,1\},\{9,3\},\{1\},\{3\},\{9\},\varnothing\}$

(f) $\{\{9,3\},\{9\},\{3\},\{1\},\varnothing\}$. These last two answers are different. The statement that $\mathbb{P}(X \cup Y) = (\mathbb{P}X) \cup (\mathbb{P}Y)$ is not always true.

(g) $\{\{3\},\varnothing\}$

(h) $\{\{3\},\varnothing\}$. These last two answers are the same. Is it always true that $\mathbb{P}(X \cap Y) = (\mathbb{P}X) \cap (\mathbb{P}Y)$?

(i) $\{\{(9,1),(3,1)\},\{(9,1)\},\{(3,1)\},\varnothing\}$

(j) $\{(\{9,3\},\{1\}),(\{9\},\{1\}),(\{3\},\{1\}),(\varnothing,\{1\}),$
$(\{9,3\},\varnothing),(\{9\},\varnothing),(\{3\},\varnothing),(\varnothing,\varnothing)\}$

Brackets have been necessary to avoid ambiguity in the order in which operations are applied; this is important as changing the order in which the operations are applied gives a different answer in general. For example $\mathbb{P}(X \times Y) \neq (\mathbb{P}X) \times (\mathbb{P}Y)$; in the first case the set product was taken before applying the power set operator, while in the second case the power set operator was applied individually to the two sets before the product operator was applied.

12. (a) i. 6

  ii. 6

  iii. 10

  iv. 12

  v. $2^2 = 4$

  vi. $2^3 = 8$

  vii. $2^2 \times 2^3 = 2^5 = 32$

  viii. $2^{2\times3} = 2^6 = 64$

  ix. $2^2 \times 2^5 = 2^7 = 128$

  Rewriting with brackets removed gives

  i. $\#E \times F$

  ii. $\#F \times E$

  iii. $\#F \times (E \cup F)$

  iv. $\#F \times (E \times F)$

  v. $\#\mathbb{P}F$

  vi. $\#\mathbb{P}E$

  vii. $\#(\mathbb{P}E) \times (\mathbb{P}F)$

  viii. $\#\mathbb{P}(E \times F)$

  ix. $\#(\mathbb{P}F) \times (\mathbb{P}(E \cup F))$

(b)   i. F

      ii. F

     iii. F

     iv. F

      v. T

     vi. T

    vii. F

   viii. T

     ix. T

      x. T

     xi. T

13.   (a) $\{(\clubsuit, \flat), \ldots\}$

     (b) $\{(\flat, \flat), \ldots\}$

     (c) $\{(\flat, \clubsuit), \ldots\}$

     (d) $\{((\clubsuit, \flat), \flat), \ldots\}$

     (e) $\{(\clubsuit, (\flat, \flat)), \ldots\}$

Using maplet notation gives

     (a) $\{\clubsuit \mapsto \flat, \ldots\}$

     (b) $\{\flat \mapsto \flat, \ldots\}$

     (c) $\{\flat \mapsto \clubsuit, \ldots\}$

     (d) $\{(\clubsuit \mapsto \flat) \mapsto \flat, \ldots\}$

     (e) $\{\clubsuit \mapsto (\flat \mapsto \flat), \ldots\}$

Notice that in the last two examples, brackets are needed to avoid ambiguity in the order in which the ordered pairs are built up.

14.   (a) 0

     (b) $2^0 = 1$

     (c) $2^1 = 2$

     (d) $2^2 = 4$. What is $\#(\mathbb{P}(\mathbb{P}(\mathbb{P}(\mathbb{P}\,\varnothing))))$?

     (e) 1

     (f) 2. What are the elements of $\{\varnothing\} \cup \{\{\varnothing\}\}$?

15.   (a) T

     (b) F

     (c) F

     (d) F

     (e) T

     (f) T

(g) F

(h) T because $\#\varnothing * X = \#\varnothing * \#X = 0 * \#X = 0$

(i) T

16. (a) T

(b) F

(c) T

(d) T

(e) F

17. (a) $\{0, 1, 3\}$

(b) $\{0, 1, 2, 3, 4\}$

(c) $\{\{0\}, \{1\}, \{0, 2\}, \{1, 2\}\}$

18. (a) $\{2, 3, 4\}$

(b) $\{2\}$

(c) $\{2, 3, 5, 6\}$

(d) $\varnothing$

(e) $\{5, 3\}$

(f) $\varnothing$

19. (a) $\#Workforce = 5$

(b) $Workforce \setminus project\_A = \{Elma, Mary\}$

(c) $project\_C \cap project\_D = \{Carlos\}$

(d) $project\_A \cup project\_B = \{Elma, Rajesh, Mary, Carlos, Mike\} = Workforce$

(e) $project\_A \times project\_C =$
$\{(Rajesh, Elma), (Rajesh, Carlos),$
$(Mike, Elma), (Mike, Carlos),$
$(Carlos, Elma), (Carlos, Carlos)\}$

(f) $project\_C \times project\_A =$
$\{(Elma, Rajesh), (Elma, Mike),$
$(Elma, Carlos), (Carlos, Rajesh),$
$(Carlos, Mike), (Carlos, Carlos)\}$

(g) $\mathbb{P} \, project\_C = \{\{Elma, Carlos\}, \{Elma\}, \{Carlos\}, \varnothing\}$

(h) $\mathbb{P}_1 \, project\_C = \{\{Elma, Carlos\}, \{Elma\}, \{Carlos\}\}$

(i) $\bigcup \{project\_A, project\_B, project\_C\} =$
$\{Elma, Rajesh, Mary, Carlos, Mike\} = Workforce$

(j) $\bigcap \{project\_A, project\_B, project\_C\} = \varnothing$

B.2   Propositional logic

1.   (a)   i.

p	$\neg p$	$\neg p \vee p$
T	F	T
F	T	T

$$\models \neg p \vee p$$

ii.

p	$\neg p$	$\neg p \wedge p$
T	F	F
F	T	F

$$\neg p \wedge p \models F$$

iii.

p	$\neg p$	$\neg \neg p$	$\neg \neg p \vee p$
T	F	T	T
F	T	F	F

$$\neg \neg p \vee p \equiv p$$

iv.

p	$\neg p$	$\neg \neg p$	$\neg \neg p \wedge p$
T	F	T	T
F	T	F	F

$$\neg \neg p \wedge p \equiv p$$

(b)   i.

p	q	$\neg p$	$\neg p \vee q$
T	T	F	T
T	F	F	F
F	T	T	T
F	F	T	T

ii.

p	q	$\neg q$	$p \wedge \neg q$
T	T	F	F
T	F	T	T
F	T	F	F
F	F	T	F

iii.

p	q	$\neg q$	$p \wedge \neg q$	$\neg(p \wedge \neg q)$
T	T	F	F	T
T	F	T	T	F
F	T	F	F	T
F	F	T	F	T

$$\neg(p \wedge \neg q) \equiv \neg p \vee q$$

iv.

p	q	$p \Rightarrow q$	$q \Rightarrow p$	$(p \Rightarrow q) \wedge (q \Rightarrow p)$
T	T	T	T	T
T	F	F	T	F
F	T	T	F	F
F	F	T	T	T

v.

p	q	$p \Rightarrow q$	$q \Rightarrow p$	$(p \Rightarrow q) \Leftrightarrow (q \Rightarrow p)$
T	T	T	T	T
T	F	F	T	F
F	T	T	F	F
F	F	T	T	T

$$(p \Rightarrow q) \wedge (q \Rightarrow p) \equiv (p \Rightarrow q) \Leftrightarrow (q \Rightarrow p)$$

vi.

p	q	$\neg p \vee p$	$\neg q \wedge q$	$(\neg p \vee p) \Rightarrow (\neg q \wedge q)$
T	T	T	F	F
T	F	T	F	F
F	T	T	F	F
F	F	T	F	F

$$(\neg p \vee p) \Rightarrow (\neg q \wedge q) \models F$$

vii.

p	q	$\neg p \wedge p$	$\neg q \wedge q$	$(\neg p \wedge p) \Rightarrow (\neg q \wedge q)$
T	T	F	F	T
T	F	F	F	T
F	T	F	F	T
F	F	F	F	T

$$\models (\neg p \wedge p) \Rightarrow (\neg q \wedge q)$$

(c)   i.

p	q	r	$p \vee q$	$(p \vee q) \vee r$
T	T	T	T	T
T	T	F	T	T
T	F	T	T	T
T	F	F	T	T
F	T	T	T	T
F	T	F	T	T
F	F	T	F	T
F	F	F	F	F

ii.

p	q	r	$q \vee r$	$p \vee (q \vee r)$
T	T	T	T	T
T	T	F	T	T
T	F	T	T	T
T	F	F	F	T
F	T	T	T	T
F	T	F	T	T
F	F	T	T	T
F	F	F	F	F

$$(p \vee q) \vee r \equiv p \vee (q \vee r)$$

We say that $\vee$ is *associative*. Note that we can write $p \vee q \vee r$ without needing to specify whether to do $p \vee q$ or $q \vee r$ first. Several other examples of this associativity follow.

iii.

p	q	r	$p \wedge q$	$(p \wedge q) \wedge r$
T	T	T	T	T
T	T	F	T	F
T	F	T	F	F
T	F	F	F	F
F	T	T	F	F
F	T	F	F	F
F	F	T	F	F
F	F	F	F	F

iv.

p	q	r	$q \wedge r$	$p \wedge (q \wedge r)$
T	T	T	T	T
T	T	F	F	F
T	F	T	F	F
T	F	F	F	F
F	T	T	T	F
F	T	F	F	F
F	F	T	F	F
F	F	F	F	F

$$(p \wedge q) \wedge r \equiv p \wedge (q \wedge r)$$

v.

p	q	r	$p \Rightarrow q$	$(p \Rightarrow q) \Rightarrow r$
T	T	T	T	T
T	T	F	T	F
T	F	T	F	T
T	F	F	F	T
F	T	T	T	T
F	T	F	T	F
F	F	T	T	T
F	F	F	T	F

vi.

p	q	r	$q \Rightarrow r$	$p \Rightarrow (q \Rightarrow r)$
T	T	T	T	T
T	T	F	F	F
T	F	T	T	T
T	F	F	T	T
F	T	T	T	T
F	T	F	F	T
F	F	T	T	T
F	F	F	T	T

$$(p \Rightarrow q) \Rightarrow r \not\equiv p \Rightarrow (q \Rightarrow r)$$

Note that $\Rightarrow$ is NOT associative. However, it is true that

$$(p \Rightarrow q) \Rightarrow r \models p \Rightarrow (q \Rightarrow r)$$

vii.

p	q	r	$p \Leftrightarrow q$	$(p \Leftrightarrow q) \Leftrightarrow r$
T	T	T	T	T
T	T	F	T	F
T	F	T	F	F
T	F	F	F	T
F	T	T	F	F
F	T	F	F	T
F	F	T	T	T
F	F	F	T	F

viii.

p	q	r	$q \Leftrightarrow r$	$p \Leftrightarrow (q \Leftrightarrow r)$
T	T	T	T	T
T	T	F	F	F
T	F	T	F	F
T	F	F	T	T
F	T	T	T	F
F	T	F	F	T
F	F	T	F	T
F	F	F	T	F

$$(p \Leftrightarrow q) \Leftrightarrow r \equiv p \Leftrightarrow (q \Leftrightarrow r)$$

ix.

p	q	r	$q \wedge r$	$p \Rightarrow (q \wedge r)$
T	T	T	T	T
T	T	F	F	F
T	F	T	F	F
T	F	F	F	F
F	T	T	T	T
F	T	F	F	T
F	F	T	F	T
F	F	F	F	T

x.

p	q	r	$p \Rightarrow q$	$p \Rightarrow r$	$(p \Rightarrow q) \wedge (p \Rightarrow r)$
T	T	T	T	T	T
T	T	F	T	F	F
T	F	T	F	T	F
T	F	F	F	F	F
F	T	T	T	T	T
F	T	F	T	T	T
F	F	T	T	T	T
F	F	F	T	T	T

$$p \Rightarrow (q \wedge r) \equiv (p \Rightarrow q) \wedge (p \Rightarrow r)$$

We say that $\Rightarrow$ is distributive over $\wedge$.

xi.

p	q	r	$p \wedge q$	$p \wedge r$	$(p \wedge q) \Rightarrow (p \wedge r)$
T	T	T	T	T	T
T	T	F	T	F	F
T	F	T	F	T	T
T	F	F	F	F	T
F	T	T	F	F	T
F	T	F	F	F	T
F	F	T	F	F	T
F	F	F	F	F	T

xii.

p	q	r	$q \vee r$	$p \Rightarrow (q \vee r)$
T	T	T	T	T
T	T	F	T	T
T	F	T	T	T
T	F	F	F	F
F	T	T	T	T
F	T	F	T	T
F	F	T	T	T
F	F	F	F	T

xiii.

p	q	r	$p \Rightarrow q$	$p \Rightarrow r$	$(p \Rightarrow q) \vee (p \Rightarrow r)$
T	T	T	T	T	T
T	T	F	T	F	T
T	F	T	F	T	T
T	F	F	F	F	F
F	T	T	T	T	T
F	T	F	T	T	T
F	F	T	T	T	T
F	F	F	T	T	T

$$p \Rightarrow (q \vee r) \equiv (p \Rightarrow q) \vee (p \Rightarrow r)$$

$\Rightarrow$ is distributive over $\vee$.

xiv.

p	q	r	$q \wedge r$	$p \Leftrightarrow (q \wedge r)$
T	T	T	T	T
T	T	F	F	F
T	F	T	F	F
T	F	F	F	F
F	T	T	T	F
F	T	F	F	T
F	F	T	F	T
F	F	F	F	T

xv.

p	q	r	$p \Leftrightarrow q$	$p \Leftrightarrow r$	$(p \Leftrightarrow q) \wedge (p \Leftrightarrow r)$
T	T	T	T	T	T
T	T	F	T	F	F
T	F	T	F	T	F
T	F	F	F	F	F
F	T	T	F	F	F
F	T	F	F	T	F
F	F	T	T	F	F
F	F	F	T	T	T

$$p \Leftrightarrow (q \wedge r) \not\equiv (p \Leftrightarrow q) \wedge (p \Leftrightarrow r)$$

⇔ is NOT distributive over ∧.

xvi.

p	q	r	$p \wedge q$	$p \wedge r$	$(p \wedge q) \Leftrightarrow (p \wedge r)$
T	T	T	T	T	T
T	T	F	T	F	F
T	F	T	F	T	F
T	F	F	F	F	T
F	T	T	F	F	T
F	T	F	F	F	T
F	F	T	F	F	T
F	F	F	F	F	T

xvii.

p	q	r	$q \vee r$	$p \Leftrightarrow (q \vee r)$
T	T	T	T	T
T	T	F	T	T
T	F	T	T	T
T	F	F	F	F
F	T	T	T	F
F	T	F	T	F
F	F	T	T	F
F	F	F	F	T

xviii.

p	q	r	$p \Leftrightarrow q$	$p \Leftrightarrow r$	$(p \Leftrightarrow q) \vee (p \Leftrightarrow r)$
T	T	T	T	T	T
T	T	F	T	F	T
T	F	T	F	T	T
T	F	F	F	F	F
F	T	T	F	F	F
F	T	F	F	T	T
F	F	T	T	F	T
F	F	F	T	T	T

xix.

p	q	r	$p \wedge \neg q$	$\neg r$	$(p \wedge \neg q) \Rightarrow \neg r$	$\neg((p \wedge \neg q) \Rightarrow \neg r)$
T	T	T	F	F	T	F
T	T	F	F	T	T	F
T	F	T	T	F	F	T
T	F	F	T	T	T	F
F	T	T	F	F	T	F
F	T	F	F	T	T	F
F	F	T	F	F	T	F
F	F	F	F	T	T	F

xx.

p	q	$p \Rightarrow q$	$\neg q \Rightarrow \neg p$	$(p \Rightarrow q) \Leftrightarrow (\neg q \Rightarrow \neg p)$
T	T	T	T	T
T	F	F	F	T
F	T	T	T	T
F	F	T	T	T

$$\models (p \Rightarrow q) \Leftrightarrow (\neg q \Rightarrow \neg p)$$

so

$$p \Rightarrow q \equiv \neg q \Rightarrow \neg p$$

xxi.

p	q	r	$p \wedge q$	$(p \wedge q) \Rightarrow r$	$p \Rightarrow r$	$q \Rightarrow r$	$(p \Rightarrow r) \vee (q \Rightarrow r)$	$((p \wedge q) \Rightarrow r) \Leftrightarrow ((p \Rightarrow r) \vee (q \Rightarrow r))$
T	T	T	T	T	T	T	T	T
T	T	F	T	F	F	F	F	T
T	F	T	F	T	T	T	T	T
T	F	F	F	T	F	T	T	T
F	T	T	F	T	T	T	T	T
F	T	F	F	T	T	F	T	T
F	F	T	F	T	T	T	T	T
F	F	F	F	T	T	T	T	T

$$\models ((p \wedge q) \Rightarrow r) \Leftrightarrow ((p \Rightarrow r) \vee (q \Rightarrow r))$$

so

$$(p \wedge q) \Rightarrow r \equiv (p \Rightarrow r) \vee (q \Rightarrow r)$$

2. (a)

p	$\neg p$	$\neg p \Rightarrow p$
T	F	T
F	T	F

$$\neg p \Rightarrow p \equiv p$$

In particular,

$$\neg p \Rightarrow p \models p$$

(b)

p	q	$p \wedge q$	$q \Rightarrow (p \wedge q)$	
T	T	T	T	*
T	F	F	T	*
F	T	F	F	
F	F	F	T	

Clearly $p \not\equiv q \Rightarrow (p \wedge q)$. However, it can be seen from the rows marked with an asterisk that $p \models q \Rightarrow (p \wedge q)$.

(c)

p	q	p ⇒ q	
T	T	T	*
T	F	F	
F	T	T	
F	F	T	

(d)

p	q	r	s	p ⇒ q	r ⇒ s	p ∨ r	q ∨ s	p ∨ r ⇒ q ∨ s	
T	T	T	T	T	T	T	T	T	*
T	T	T	F	T	F	T	T	T	
T	T	F	T	T	T	T	T	T	*
T	T	F	F	T	T	T	T	T	*
T	F	T	T	F	T	T	T	T	
T	F	T	F	F	F	T	F	F	
T	F	F	T	F	T	T	T	T	
T	F	F	F	F	T	T	F	F	
F	T	T	T	T	T	T	T	T	*
F	T	T	F	T	F	T	T	T	
F	T	F	T	T	T	F	T	T	*
F	T	F	F	T	T	F	T	T	*
F	F	T	T	T	T	T	T	T	*
F	F	T	F	T	F	T	F	F	
F	F	F	T	T	T	F	T	T	*
F	F	F	F	T	T	F	F	T	*

B.3   Predicate logic

1. $1, -1$ are the only elements of the set.
2. $\varnothing$, $\{1\}$ and $\{1001\}$.
3. $(1, 2), (3, -1)$ are the only elements of the set.
4. (a) $\{1, 2, 3, 4\}$

   (b) $\{0, 1, 2, 3, 4\}$

   (c) $\{5, 6\}$

   (d) $\{0, 3, 6, 9\}$

   (e) $\{0, 1, 4\}$

   (f) $\{\{1\}, \{2\}, \{3\}\}$

   (g) $\{(-2, 4), (-1, 1), (0, 0), (1, 1), (2, 4)\}$

   (h) $\{(0, 0), (1, 0), (2, 0), (3, 0)\}$

    (i) $\{(0,0),(1,1),(2,2),(3,3)\}$ which is the same as $\{x :\in \mathbb{N} \mid x < 4 \bullet (x,x)\}$

    (j) $\{(2,1)\}$

5. (a) y is free

    (b) x is bound (by the set comprehension)

    (c) y is bound (by the set comprehension) but x is free

6. For each question there are many possible solutions, but some typical answers include:

    (a) $\{y :\in \mathbb{N} \mid y < 4 \bullet 4 * y\}$

    (b) $\{z :\in \{0,1,2\} \bullet \{z\}\}$ or $\{x :\in \mathbb{PN} \mid \#x = 1 \wedge (\forall y :\in x \bullet y < 3)\}$

    (c) $\{z :\in \{0,1,2\} \bullet \{3 * z\}\}$

    (d) $\{w :\in \mathbb{N} \mid w \leqslant 4 \bullet (w, 2 * w)\}$ or $\{x :\in \mathbb{N} \times \mathbb{N} \mid x.1 < 5 \wedge x.2 = 2 * x.1\}$

    (e) $\{y :\in \mathbb{N} \mid y < 5 \bullet (2 * y, 2 * y)\}$

7. (a) $\{x :\in \mathbb{Z} \mid x > 10\} \cup \{x :\in \mathbb{Z} \mid x < 8\}$

    (b) $\{x :\in \mathbb{Z} \mid x > 6\} \cap \{x :\in \mathbb{Z} \mid x < 8\}$

    (c) $\{x :\in \mathbb{Z} \mid x \notin \mathbb{N}\} \cup \{x :\in \mathbb{Z} \mid x < 8\}$

    (d) $\{x :\in \mathbb{Z} \mid x \in \mathbb{N}\} \cap \{x :\in \mathbb{Z} \mid x \geqslant 8\}$ which in fact is simply $\{x :\in \mathbb{Z} \mid x \geqslant 8\}$.

8. (a) $\{x :\in \mathbb{Z} \mid 6 < x < 7\}$ which is in fact $\varnothing$

    (b) $\{x :\in \mathbb{Z} \mid x > 6 \vee x < 9\}$ which is in fact $\mathbb{Z}$

    (c) $\{x :\in \mathbb{Z} \mid \neg\, 0 < x \leqslant 6\}$

9. $0^2 + 1^2 + 2^2 = 5$

10. (a)

x	$x^2 > 4$	TruthValue
0	$0^2 > 4$	F
1	$1^2 > 4$	F
2	$2^2 > 4$	F
3	$3^2 > 4$	T
4	$4^2 > 4$	T

    (b)

y	$y < 20$	TruthValue
0	$0 < 20$	T
1	$1 < 20$	T
2	$2 < 20$	T
3	$3 < 20$	T
4	$4 < 20$	T

(c)

x	z	$z^2 > 4 \Rightarrow x * z > 4$	TruthValue
0	1	$1^2 > 4 \Rightarrow 0 * 1 > 4$	T
0	2	$2^2 > 4 \Rightarrow 0 * 2 > 4$	T
0	3	$3^2 > 4 \Rightarrow 0 * 3 > 4$	F
1	1	$1^2 > 4 \Rightarrow 1 * 1 > 4$	T
1	2	$2^2 > 4 \Rightarrow 1 * 2 > 4$	T
1	3	$3^2 > 4 \Rightarrow 1 * 3 > 4$	F

(d) Although the possible values of y include 0, there are no values of w in $\mathbb{N}$ for which $w < 0$:

y	w	$w^2 > 4 \Leftrightarrow \neg\, y^2 > 4$	TruthValue
1	0	$0^2 > 4 \Leftrightarrow \neg\, 1^2 > 4$	F
2	0	$0^2 > 4 \Leftrightarrow \neg\, 2^2 > 4$	F
2	1	$1^2 > 4 \Leftrightarrow \neg\, 2^2 > 4$	F

(e) A few sample values for y are

y	$\#y = 5$	TruthValue
$\{\}$	$\#\{\} = 5$	F
$\{0, 1, 2, 3\}$	$\#\{0, 1, 2, 3\} = 5$	F
$\{2, 3, 5, 7, 11\}$	$\#\{2, 3, 5, 7, 11\} = 5$	T

(f) Changing the free variable from y to z makes no essential difference.

z	$\#z = 5$	TruthValue
$\{\}$	$\#\{\} = 5$	F
$\{0, 1, 2, 3\}$	$\#\{0, 1, 2, 3\} = 5$	F
$\{2, 3, 5, 7, 11\}$	$\#\{2, 3, 5, 7, 11\} = 5$	T

(g) A few sample values for x and y are

x	y	$\#y = x$	TruthValue
0	$\varnothing$	$\#\varnothing = 0$	T
3	$\{0, 1, 2, 3\}$	$\#\{0, 1, 2, 3\} = 3$	F
5	$\{2, 3, 5, 7, 11\}$	$\#\{2, 3, 5, 7, 11\} = 5$	T

11. (a) F

(b) T

(c) T

(d) T

(e) F

(f) T

(g) T

(h) T

(i) F

(j) T

(k) *F*

12.   (a) $\{3,6\} = \{\, x :\in \mathbb{N}_1 \mid x < 3 \bullet 3 * x \,\}$

(b) $\{0,1\} = \{\, x :\in \mathbb{N} \mid x < 2 \,\}$

(c) $\{1,2,3\} = \{\, x :\in \mathbb{N}_1 \mid x < 4 \,\}$

(d) $\{6\} = \{\, w :\in \mathbb{Z} \mid w = 6 \,\}$

(e) $\{\{0\}, \{1\}, \{2\}, \{0,1\}, \{0,2\}, \{1,2\}, \{0,1,2\}\}$
$= \{\, z :\in \mathbb{P}\{0,1,2\} \mid z \neq \varnothing \,\}$

13.   (a) *Mike* $\in$ *new_project*.

(b) *#new_project* $\geqslant 2$.

(c) *Elma* $\in$ *new_project* $\Rightarrow$ *Rajesh* $\in$ *new_project*.

(d) $\exists$ *member* $:\in$ *new_project* $\bullet$ *member* $\in$ *project_A*.

Alternatively $\exists$ *member* $:\in$ *project_A* $\bullet$ *member* $\in$ *new_project*; although this seems to be saying something different, the end result is the same! This duality may be a little confusing: which set do we choose to give us the possible values of *member* and which to go into the predicate? The best way round this problem is allow *member* to take values from the biggest set possible, in this case *Workforce*, and modify the predicate accordingly. In this case we would write:

$$\exists \text{ member} :\in \text{Workforce} \bullet \text{member} \in \text{new\_project} \cap \text{project\_A}$$

Yet another alternative, and arguably the best, is

$$\text{new\_project} \cap \text{project\_A} \neq \varnothing$$

(e) $\forall$ *member* $:\in$ *Workforce* $\bullet$
member $\in$ *new_project* $\Rightarrow$ *member* $\in$ *project_A*.
Alternatively (and more simply) we can write *new_project* $\subseteq$ *project_A*.

(f) $\forall$ *member* $:\in$ *Workforce* $\bullet$
member $\in$ *project_C* $\Rightarrow$ *member* $\in$ *new_project*.
Alternatively, *project_C* $\subseteq$ *new_project*.

(g) ($\exists$ *member* $:\in$ *Workforce* $\bullet$ *member* $\in$ *new_project* $\cap$ *project_C*) $\vee$
($\forall$ *member* $:\in$ *Workforce* $\bullet$
member $\in$ *project_D* $\Rightarrow$ *member* $\in$ *new_project*).
Alternatively,
new_project $\cap$ *project_C* $\neq \varnothing \vee$ *project_D* $\subseteq$ *new_project*.

(h) $\forall$ *member* $:\in$ *Workforce* $\bullet$
(*member* $\in$ *new_project* $\Rightarrow$ *member* $\in$ *project_A* $\wedge$
member $\in$ *new_project* $\Rightarrow$ *member* $\in$ *project_B*).
Alternatively,
$\forall$ *member* $:\in$ *Workforce* $\bullet$
member $\in$ *new_project* $\Rightarrow$ (*member* $\in$ *project_A* $\cap$ *project_B*).
Yet another alternative is simply *new_project* $\subseteq$ *project_A* $\cap$ *project_B*.

(i) ∀ *member* :∈ *Workforce* •
   (*member* ∈ *new_project* ⇒ *member* ∈ *project_A* ∨
   member ∈ *new_project* ⇒ *member* ∈ *project_B*).
   Alternatively,
   ∀ *member* :∈ *Workforce* •
   member ∈ *new_project* ⇒ (*member* ∈ *project_A* ∪ *project_B*).
   Yet another alternative is simply *new_project* ⊆ *project_A* ∪ *project_B*.

B.4 Relations

1. (a) $I_X = \{(0,0),(1,1),(2,2),(3,3)\}$
   (b) $I_Y = \{(4,4),(5,5),(6,6)\}$
   (c) $I_Z = \{(7,7),(8,8)\}$
   (d) $\operatorname{dom} R = \{0,1,3\}$
   (e) $\operatorname{ran} R = \{4,5,6\}$
   (f) $\operatorname{dom} S = \{0,1,2,3\}$
   (g) $\operatorname{ran} S = \{7,8\}$
   (h) $\operatorname{dom} T = \{4,5\}$
   (i) $\operatorname{ran} T = \{7,8\}$
   (j) $\operatorname{dom} U = \{4,5,6\}$
   (k) $\operatorname{ran} U = \{2\}$
   (l) $\operatorname{dom} V = \{7\}$
   (m) $\operatorname{ran} V = \{8\}$
   (n) $R^\sim = \{(4,0),(5,0),(5,1),(6,1),(4,3)\}$
   (o) $S^\sim = \{(7,0),(7,1),(7,2),(7,3),(8,2)\}$
   (p) $T^\sim = \{(7,4),(8,4),(7,5)\}$
   (q) $U^\sim = \{(2,4),(2,5),(2,6)\}$
   (r) $V^\sim = \{(8,7)\}$
2. (a) $R \,\mathring{,}\, I_Y = R$
   (b) $I_X \,\mathring{,}\, R = R$
   (c) $R \,\mathring{,}\, T = \{(0,7),(0,8),(1,7),(3,7),(3,8)\}$
   (d) $T \,\mathring{,}\, V = \{(4,8),(5,8)\}$
   (e) $S \,\mathring{,}\, V = \{(0,8),(1,8),(2,8),(3,8)\}$
   (f) $R \,\mathring{,}\, U = \{(0,2),(1,2),(3,2)\}$

(g) $(R \,\overset{\circ}{\circ}\, T) \,\overset{\circ}{\circ}\, V = \{(0,8),(1,8),(3,8)\}$

(h) $R \,\overset{\circ}{\circ}\, (T \,\overset{\circ}{\circ}\, V) = (R \,\overset{\circ}{\circ}\, T) \,\overset{\circ}{\circ}\, V$

(i) $V \,\overset{\circ}{\circ}\, V = \{\}$

3. (a) $R \,\overset{\circ}{\circ}\, R^{\sim} = \{(0,0),(0,3),(0,1),(1,0),(1,1),(3,0),(3,3)\} \subseteq (\operatorname{dom} R)^2$

(b) $R^{\sim} \,\overset{\circ}{\circ}\, R = \{(4,4),(4,5),(5,4),(5,5),(5,6),(6,5),(6,6)\} \subseteq (\operatorname{ran} R)^2$

(c) $S \,\overset{\circ}{\circ}\, S^{\sim} = X^2 = (\operatorname{dom} S)^2$

(d) $S^{\sim} \,\overset{\circ}{\circ}\, S = Z^2 = (\operatorname{ran} S)^2$

(e) $T \,\overset{\circ}{\circ}\, T^{\sim} = \{(4,4),(4,5),(5,4),(5,5)\} = (\operatorname{dom} T)^2$

(f) $T^{\sim} \,\overset{\circ}{\circ}\, T = Z^2 = (\operatorname{ran} T)^2$

(g) $U \,\overset{\circ}{\circ}\, U^{\sim} = Y^2 = (\operatorname{dom} U)^2$

(h) $U^{\sim} \,\overset{\circ}{\circ}\, U = \{(2,2)\} = (\operatorname{ran} U)^2$

(i) $V \,\overset{\circ}{\circ}\, V^{\sim} = \{(7,7)\} = (\operatorname{dom} V)^2$

(j) $V^{\sim} \,\overset{\circ}{\circ}\, V = \{(8,8)\} = (\operatorname{ran} V)^2$

(k) $(R \,\overset{\circ}{\circ}\, T) \,\overset{\circ}{\circ}\, S^{\sim} = \{(0,0),(0,1),(0,2),(0,3),(1,0),(1,1),(1,2),(1,3),$
$(3,0),(3,1),(3,2),(3,3)\}$

(l) $R \,\overset{\circ}{\circ}\, (T \,\overset{\circ}{\circ}\, S^{\sim}) = (R \,\overset{\circ}{\circ}\, T) \,\overset{\circ}{\circ}\, S^{\sim}$

4. (a) *accommodates* =
    $\{(Big, Elma), (Big, Rajesh), (Big, Mary), (Little, Carlos), (Little, Mike)\}$
    represents the allocation of people to offices.

(b) dom *accommodates* = $\{Big, Little\}$ represents the current offices in use.

(c) ran *accommodates* = $\{Elma, Rajesh, Mary, Carlos, Mike\}$ represents
    the current workforce.

(d) *accommodates*$^{\sim}$ =
    $\{(Elma, Big), (Rajesh, Big), (Mary, Big), (Carlos, Little), (Mike, Little)\}$
    represents the allocation of offices to people.

(e) *accommodates* $\,\overset{\circ}{\circ}\,$ *accommodates* = $\{(Big, Big), (Little, Little)\}$ repre-
    sents offices which have an occupant in common. In this case we can
    see that no-one has more than one office.

(f) *accommodates*$^{\sim}$ $\,\overset{\circ}{\circ}\,$ *accommodates*$^{\sim}$ =
    $\{(Elma, Elma), (Elma, Rajesh), (Elma, Mary),$
    $(Rajesh, Elma), (Rajesh, Rajesh), (Rajesh, Mary),$
    $(Mary, Elma), (Mary, Rajesh), (Mary, Mary),$
    $(Mike, Mike), (Mike, Carlos), (Carlos, Mike), (Carlos, Carlos)\}$
    represents the relation of sharing an office.

(g) *accommodates* $\,\overset{\circ}{\circ}\,$ *project_member* =
    $\{(Big, A), (Big, B), (Big, C), (Big, D),$
    $\{(Little, A), (Little, B), (Little, C), (Little, D)\}$
    represents the relation between offices and projects worked upon by
    people from the offices.

5. *Allocations =*
   $\{(Big, Elma, B)\,, (Big, Elma, C)\,,$
   $(Big, Rajesh, A)\,, (Big, Rajesh, B)\,, (Big, Rajesh, C)\,,$
   $(Big, Mary, B)\,, (Big, Mary, D)\,,$
   $(Little, Carlos, A)\,, (Little, Carlos, C)\,, (Little, Carlos, D)\,,$
   $(Little, Mike, A)\,, (Little, Mike, B)\,, (Little, Mike, D)\}$
   Although *Allocations* has fewer elements (13) than the total of the two binary relations, each element is more complex. Furthermore, the same information is stored more than once. For example the fact that *Elma* is in the *Big* office is stored in each of the first two elements listed above. Changing the value of the one complex relation will be more complicated than changing the value of one binary relation.

B.5   Homogeneous relations

The following abbreviations are used: 'r' means reflexive; 's' means symmetric; 'a' means antisymmetric; 't' means transitive. So, for example, an equivalence relation would have the description 'rst'.

1. (a) rat   (c) s   (e) rst
   (b) None   (d) t $\{0\}, \{1,2\}$

2. (a) rsat $\{0\}, \{1\}, \{2\}, \{3\}$ (f) sat   (k) rs
   (b) s   (g) sat   (l) st
   (c) rst Y (h) rst Z (m) rst $\{0,1,3\}, \{2\}$
   (d) st   (i) rst Z
   (e) rst X (j) at   (n) sat

3. Just one, namely the identity relation, I_A.
4. The symmetric closures are

   (a) $\{(0,0)\,, (0,2)\,, (0,4)\,, (2,0)\,, (4,0)\}$
   (b) $\{(2,0)\,, (0,2)\}$ is already symmetric, and hence equal to its symmetric closure.
   (c) $\{(0,0)\,, (2,4)\,, (4,2)\}$
   (d) $\{(2,4)\,, (4,2)\,, (2,0)\,, (0,2)\}$

   The transitive closures are

   (a) $\{(0,0)\,, (0,2)\,, (0,4)\}$ is already transitive and hence equal to its transitive closure.

(b) $\{(2,0),(0,2),(0,0),(2,2)\}$

(c) $\{(0,0),(2,4)\}$ is already transitive and hence equal to its transitive closure.

(d) $\{(2,4),(4,2),(2,0),(2,2),(4,4),(4,0)\}$

B.6 Functions

1. The following sets are functions: a, c, d, e, f, g, j, l, n.
2. Domain and range are (respectively in each case)

   (a) $\{0,1,2,7\}$ and $\{3,4,5,7\}$

   (b) $\{0,1,2,3\}$ and $\{0,3,4,7\}$

   (c) $\{1,2,3,7\}$ and $\{3\}$

   (d) $\{\}$ and $\{\}$

   (e) $\{(0,2),(3,5)\}$ and $\{(6,4),(7,1)\}$

   (f) $\{\{2,3\},\{1,2\},\{1,3\}\}$ and $\{\{1\},\{2\},\{3\}\}$

   (g) $\{\{\}\}$ and $\{\{\}\}$

   (h) $\{Elma, Mike, Rajesh\}$ and $\{A, C, D\}$

   (i) $\{Elma, Carlos, Mike, Rajesh\}$ and $\{A, B, C, D\}$

   (j) $\{Elma, Carlos, Mike, Mary\}$ and $\{A, B, C, D\}$

   (k) $\{Elma, Carlos, Mike, Mary, Rajesh\}$ and $\{A, B, C, D\}$

   (l) $\{Elma, Carlos, Mike\}$ and $\{Big\}$

   (m) $\{(Big, Elma),(Big, Rajesh),(Little, Elma)\}$ and $\{A, B, C\}$

3. The following functions are:

   - total – a, c, d, i, j, m;
   - injective – a, b, c, e, g, h, i, j, k, l;
   - surjective – b, c, h, j, l, m, o;
   - bijective – c, j (these are the only two to occur in all three of the above lists).

4. **(a)** $\{(4,2),(3,0),(7,1),(5,7)\}$ is a function

   (b) $\{(4,2),(3,0),(7,1),(0,3)\}$ is a function

   (c) $\{(3,1),(3,2),(3,3),(3,7)\}$ is not a function

   (d) $\{\}$ is a function

   (e) $\{((6,4),(0,2)),((7,1),(3,5))\}$ is a function

(f) $\{(\{1\}, \{2, 3\}), (\{3\}, \{1, 2\}), (\{2\}, \{1, 3\})\}$ is a function

(g) $\{(\{\}, \{\})\}$ is a function

(h) $\{(A, Elma), (B, Carlos), (C, Mike)\}$ is a function

(i) $\{(A, Elma), (B, Carlos), (C, Mike), (D, Mary)\}$ is a function

(j) $\{(A, Elma), (B, Carlos), (C, Mike), (D, Mary), (D, Rajesh)\}$ is not a function

(k) $\{(Big, Elma), (Big, Carlos), (Big, Mike)\}$ is not a function

(l) $\{(Big, Elma), (Big, Carlos), (Little, Mike)\}$ is not a function

5. (a) $\{4 \mapsto 2, 0 \mapsto 4, 2 \mapsto 4, 3 \mapsto 3, 5 \mapsto 5\}$

(b) $\{4 \mapsto 2, 0 \mapsto 4, 2 \mapsto 4, 3 \mapsto 3, 5 \mapsto 3\}$

(c) $\{0 \mapsto 4, 2 \mapsto 4, 3 \mapsto 3, 5 \mapsto 5\}$

(d) $\{0 \mapsto 4, 2 \mapsto 4, 3 \mapsto 3, 5 \mapsto 5\}$

(e) $\{1 \mapsto 5\}$

(f) $\{\}$

(g) $\{\}$

(h) $\{2 \mapsto 2, 0 \mapsto 0, 1 \mapsto 1, 7 \mapsto 7\}$

(i) $\{2 \mapsto 2, 0 \mapsto 0, 1 \mapsto 1, 3 \mapsto 3\}$

(j) $\{4 \mapsto 4, 3 \mapsto 3, 7 \mapsto 7, 5 \mapsto 5\}$

(k) $\{4 \mapsto 4, 3 \mapsto 3, 7 \mapsto 7, 0 \mapsto 0\}$

(l) $\{1 \mapsto 2, 7 \mapsto 0, 5 \mapsto 1\}$

(m) $\{7 \mapsto 4, 5 \mapsto 3\}$

(n) $\{1 \mapsto 2, 7 \mapsto 0, 5 \mapsto 1\}$

6. (a) $\{4 \mapsto 1, 0 \mapsto 3, 2 \mapsto 3, 3 \mapsto 2, 5 \mapsto 4, 1 \mapsto 2\}$

(b) $\{1 \mapsto 2, 2 \mapsto 3, 3 \mapsto 4, 4 \mapsto 5, 0 \mapsto 3, 5 \mapsto 4\}$

(c) $\{4 \mapsto 1, 0 \mapsto 3, 2 \mapsto 3, 3 \mapsto 2, 5 \mapsto 4\}$

(d) $\{4 \mapsto 1, 0 \mapsto 3, 2 \mapsto 3, 3 \mapsto 2, 5 \mapsto 4\}$

(e) $\{4 \mapsto 1, 0 \mapsto 3, 2 \mapsto 3, 3 \mapsto 2, 5 \mapsto 4\}$

(f) $\{4 \mapsto 1, 0 \mapsto 3, 2 \mapsto 3, 3 \mapsto 2, 5 \mapsto 4\}$

(g) $\{4 \mapsto 1, 0 \mapsto 3, 2 \mapsto 3, 3 \mapsto 2, 5 \mapsto 4\}$

(h) $\{2 \mapsto 4, 0 \mapsto 3, 1 \mapsto 7, 7 \mapsto 5, 4 \mapsto 2, 3 \mapsto 0, 5 \mapsto 7\}$

(i) $\{(2, 4), (0, 3), (1, 7), (7, 5), (4, 1), (3, 7)\}^{\sim}$
$= \{(4, 2), (3, 0), (7, 1), (5, 7), (1, 4), (7, 3)\}$ which is *not* a function.
Note that in the question the '$\sim$' notation was used; the '$^{-1}$' notation is not appropriate since the override of two injective functions is not necessarily injective.

(j) $\{4 \mapsto 2, 3 \mapsto 0, 7 \mapsto 1, 5 \mapsto 7, 1 \mapsto 4\}$

(k) $\{1 \mapsto 4, 7 \mapsto 3, 5 \mapsto 7, 4 \mapsto 2, 3 \mapsto 0\}$

7. (a) $\{(Elma, A), (Carlos, B), (Mike, C), (Mary, D)\}$

(b) $\{(A, Big), (B, Big), (C, Big)\}$

(c) $\{(A, Big), (B, Big), (C, Little)\}$

B.7   Applications

1. (a) $S_1(3) = a$

(b) $\operatorname{dom} S_1 = \{1, 2, 3, 4, 5\}$

(c) $\operatorname{ran} S_2 = \{b, c\}$

2. (a) $S_1 = \{1 \mapsto a, 2 \mapsto b, 3 \mapsto a, 4 \mapsto a, 5 \mapsto c\} = \langle a, b, a, a, c \rangle$

(b) $S_1{}^{\sim} = \{(a, 1), (b, 2), (a, 3), (a, 4), (c, 5)\}$

(c) $S_1 \oplus S_2 = \{1 \mapsto c, 2 \mapsto b, 3 \mapsto c, 4 \mapsto a, 5 \mapsto c\} = \langle c, b, c, a, c \rangle$
The operation updates a sublist at the start.

(d) $S_2 \oplus S_1 = \{1 \mapsto a, 2 \mapsto b, 3 \mapsto a, 4 \mapsto a, 5 \mapsto c\} = \langle a, b, a, a, c \rangle$

(e) $S_1 \cup S_2 = \{(1, a), (2, b), (3, a), (4, a), (5, c), (1, c), (2, b), (3, c)\}$

(f) $S_1 \cap S_2 = \{2 \mapsto b\}$. The operation in itself does not achieve anything
particularly useful, but $\operatorname{ran}(S_1 \cap S_2)$ gives the set of items which share
common positions in the two lists, while $\operatorname{dom}(S_1 \cap S_2)$ is the set of
positions in the two lists which have items in common.

(g) $S_1 \, {}^\circ_9 \, S_2{}^{\sim} = \{(2, 2), (5, 1), (5, 3)\}$

(h) $S_1 \, {}^\circ_9 \, S_2{}^{\sim} \, {}^\circ_9 \, S_1 = \{2 \mapsto b, 5 \mapsto a\}$

(i) $S_2 \, {}^\circ_9 \, S_2{}^{\sim} = \{(1, 1), (1, 3), (3, 1), (3, 3), (2, 2)\}$
This is an equivalence relation on $\{1, 2, 3\}$ with equivalence classes of
$\{1, 3\}$, the positions of c in the list, and $\{2\}$, the position of b in the list.

(j) $S_2 \, {}^\circ_9 \, S_2{}^{\sim} \, {}^\circ_9 \, S_2 = \{1 \mapsto c, 2 \mapsto b, 3 \mapsto c\} = \langle c, b, c \rangle$
In general for *any* sequence S it is true that $S \, {}^\circ_9 \, S^{\sim} \, {}^\circ_9 \, S = S$.

3. $\mathbb{P} S_2 = \{\{1 \mapsto c, 2 \mapsto b, 3 \mapsto c\}, \{2 \mapsto b, 3 \mapsto c\}, \{1 \mapsto c, 3 \mapsto c\}, \{1 \mapsto c, 2 \mapsto b\}, \{1 \mapsto c\}, \{2 \mapsto b\}, \{3 \mapsto c\}, \{\}\}$
There are four sequences, all subsequences of S_2. They are $\langle c, b, c \rangle$, $\langle c, b \rangle$,
$\langle c \rangle$, $\langle \rangle$. Note the empty sequence $\langle \rangle$, which is modelled by the empty set. The
definition given in the question is rather imprecise, since it could be interpreted
that the empty set does not correspond to a sequence. The imprecision arises
because we have not formally defined the meaning of $1 \ldots n$:

$$1 \ldots n = \{x : \in \mathbb{N} \mid 1 \leqslant x \leqslant n\}$$

With this definition, $1 \ldots 0 = \varnothing$. Thus the empty set does indeed represent a sequence.

4. A possible definition for $\frown$ in terms of operations on sets, relations, and functions is:

$$S \frown T = S \cup \{\, x :\in \operatorname{dom} T \bullet (x + \#S) \mapsto T(x) \,\}$$

In this definition, the position (x) of each element in T is increased by the length $(\#S)$ of S so that the sequence T is appended to the sequence S.

B.8   Self-test questions

CADED BEDEA ECCCD DEAEE DBDDC

Glossary of terms

1–1 correspondence See **Bijection**.

AND See **Conjunction**.

Biconditional A **Logical Connective** denoted by $\Leftrightarrow$. The expression $p \Leftrightarrow q$ is T whenever p and q have the same truth values, but F otherwise.

Bijection A bijection is a **Total function** which is both an **Injection** and a **Surjection**.

Bijective function See **Bijection**.

Bound variable A variable, x say, which is bound by a **Set comprehension** or a quantifier, such as a summation, a **Universal quantifier**, $\forall$ or an **Existential quantifier**, $\exists$.

Cardinality The cardinality of a set A, denoted $\#A$, is an integer $(0, 1, 2 \ldots)$; the cardinality of a finite set is interpreted as the number of elements in the set. More generally, sets A and B have the same cardinality if a **Bijection** can be found between them.

Cartesian product of sets See **Product of sets**.

Codomain An alternative name for target set. See **Relation**.

Complement See **Difference of sets**.

Composition of functions The composition of functions g and f is itself a function, denoted by $g \circ f$. The expression $g \circ f(x)$, more usually written $g(f(x))$, is the result of first applying f to the value x, and then applying g to the result of $f(x)$. Regarding functions as a special kind of **Relation**, the composition of functions is a particular instance of relational composition. In particular $g \circ f \equiv g \, {}_9^8 f$.

Composition of relations The composition of relations R and S is another relation denoted by $R \, {}_9^8 S$; two objects x and z are related under this composite relation if and only if there is at least one object y for which x is related to y under R and y is related to z under S. In terms of **Set**s, $(x, z) \in R \, {}_9^8 S$ if and only if $(x, y) \in R$ and $(y, z) \in S$.

Conditional A **Logical Connective** denoted by $\Rightarrow$. The expression $p \Rightarrow q$ is always T except when p is T and q is F.

Conjunction A **Logical Connective** denoted by $\wedge$. The expression $p \wedge q$ is only T when both p and q are T.

Connective See **Logical Connective**.

Contradiction A **Propositional form** which is always F no matter what propositions are substituted for the letters in the propositional form. See also **Tautology**. A contradiction P can be denoted by $F \models P$.

Difference of sets The difference of sets A and B, written as $A \setminus B$ or as $A - B$, is obtained by removing from A all those elements which are also in B. The set difference $A \setminus B$ is often called the complement of B with respect to A.

Disjunction A **Logical Connective** denoted by $\vee$. The expression $p \vee q$ is only F when both p and q are F.

Domain The domain of a **Relation** or a **Function** is the set of all the first coordinates of the **Ordered pairs** which comprise that **Relation** or **Function**. See also **Range**.

Element See **Set**.

Empty set The empty set, $\varnothing$, is the unique set which has no members; the **Set enumeration** is $\{\}$.

Equality of sets Sets A and B are equal, $A = B$, if and only if A is a subset of B and also B is a subset of A, that is both $A \subseteq B$ and $B \subseteq A$ are true.

Equivalence Two **Propositional forms** P and Q are equivalent if they always have the same truth values no matter what **Propositions** are substituted for the letters. We denote equivalence by writing $P \equiv Q$.

Equivalence relation A **Homogeneous relation** which is a **Reflexive relation**, a **Symmetric relation** and a **Transitive relation**.

Existential quantifier The symbol $\exists$, which binds the variable to which it is applied (making the variable a **Bound variable**); the **Proposition** $\exists x :\in A \bullet p(x)$ is T if there is at least one member of the set A for which $p(x)$ is T. If $p(x)$ is F for every element of the set A, then $\exists x :\in A \bullet p(x)$ is F also.

Free variable A variable, x say, is free if it is not a **Bound variable**.

Function In terms of set theory, a function is regarded as a special type of **Relation** in which each element of the **Domain** is mapped to just one element in the **Range**.

Generalized Intersection The generalized intersection of a set of sets, denoted by $\cap\{S_1, S_2, S_3, \ldots, S_n\}$, is obtained by repeated application of **Intersection of sets**, that is, $S_1 \cap S_2 \cap S_3 \cap \ldots \cap S_n$.

Generalized Union The generalized union of a set of sets, $\cup\{S_1, S_2, S_3, \ldots, S_n\}$, is obtained by repeated application of **Union of sets**, $S_1 \cup S_2 \cup S_3 \cup \ldots \cup S_n$.

Homogeneous relation A **Relation** for which the **Source set** and **Target set** are the same.

Identity relation The **Homogeneous relation** in which every element of the source set is related to itself. For the set X, the identity relation is given by the equation $I_X = \{ x :\in X \bullet (x, x) \}$.

IF AND ONLY IF See **Biconditional**.

Injection A **Function** f is an injection if its **Relational inverse**, $f^{\sim}$, is also a function (the **Inverse function**).

Injective function See **Injection**.

Intersection of sets The intersection of sets A and B, denoted by $A \cap B$, is the set of elements in both A and B.

Inverse See **Relational inverse** and **Inverse function**.

Inverse function The **Relational inverse** of an **Injective function**, f, is also a function, called the inverse of f and denoted by f^{-1}.

Logical Connective **Proposition**s can be combined to form a compound proposition by the use of one or more logical connectives. For further information see under the various connectives: **Biconditional, Conjunction, Conditional, Disjunction** and **Negation**.

Logical implication Propositional forms $P_1, P_2, P_3, \ldots$ logically imply propositional form Q, written $P_1, P_2, P_3, \ldots \models Q$, if whenever $P_1, P_2, P_3, \ldots$ is T then Q is also T.

Maplet notation See **Ordered pair**.

Material conditional See **Conditional**.

Member See **Set**.

Negation A **Logical Connective** denoted by $\neg$. The expression $\neg p$ has the opposite truth value to p.

NOT See **Negation**.

Null set See **Empty set**.

ONLY IF See **Conditional**.

OR See **Disjunction**.

Ordered pair An ordered pair is a list of two elements, called coordinates, enclosed between parentheses. For example the ordered pair (x, y) has first coordinate x and second coordinate y. Sometimes the maplet notation $x \mapsto y$ is used to denote the ordered pair (x, y).

Override The override of relation R by relation S, denoted by $R \oplus S$, is obtained by adding to S all those **Ordered pairs** of R whose first coordinates are not in the **Domain** of S.

Power set The power set of set B is the set of all **Subsets** of B; it is written $\mathbb{P}B$. Set A is a subset of B if and only if it is an element of $\mathbb{P}B$; that is $A \subseteq B$ if and only if $A \in \mathbb{P}B$.

Predicate A logical expression in which there are one or more **Free variables**. Replacing all the free variables by constant values results in a **Proposition**.

Product of sets The (Cartesian or direct) product of sets A and B, denoted by $A \times B$, is the set of all possible **Ordered pairs** with first coordinate taken from A and second coordinate taken from B.

Proper subset A proper subset of set B is a **Subset** which is not equal to B. If A is a proper subset of B then we write $A \subset B$.

Proposition A proposition is a statement which is either true (T) or false (F).

Propositional form A logical expression comprising **Connectives** and letters (usually p, q, r etc.); replacing the letters by propositions gives a compound proposition.

Range The range of a **Relation** or a **Function** is the set of all the second coordinates of the **Ordered pairs** which comprise that **Relation** or **Function**. See also **Domain**.

Reflexive closure The reflexive closure of a **Homogeneous relation**, R, on a set X is the **Union of sets** $R \cup I_X$, where I_X is the **Identity relation** on X. It is a reflexive relation.

Reflexive relation A **Homogeneous relation** in which every element of the **Source set** is related to itself. See also **Reflexive relation**, **Symmetric relation**, **Transitive relation** and **Equivalence relation**.

Relation A relation between a source set A and a target set (or codomain) B is a subset of the product $A \times B$ of A with B. A **Function** is considered to be a special kind of relation.

Relational inverse The inverse of a **Relation** R is denoted by $R^{\sim}$ and is obtained by reversing the order of all the **Ordered Pairs** in R (i.e. all the second coordinates become first coordinates, and all the first coordinates become second coordinates).

Relative complement See **Difference of sets**.

Set Although a set cannot be defined, a useful approximation is to think of a set as being associated with a procedure which determines whether or not a given object is a **Member** (or **Element**) of that set. If x is a member of set A then we write $x \in A$.

Set comprehension A representation for a set in which a **Predicate** is used as the criterion for determining whether or not any given object is a member of the set.

Set display See **Set enumeration**.

Set enumeration A set enumeration consists of a list of items enclosed in braces, $\{\ldots\}$; a given object is a member of the set if it matches at least one of the items in the list.

Source set See **Relation**.

Subset A set A is a subset of set B if every element of A is also an element of B; we write $A \subseteq B$. Note that every set has both itself and the **Empty set** as subsets: $B \subseteq B$ and $\varnothing \subseteq B$. See also **Power set, Proper subset**.

Surjection A **Function** f is a surjection if its **Range** $\operatorname{ran} f$ is equal to its **Target set**.

Surjective function See **Surjection**.

Symmetric closure The reflexive closure of a **Homogeneous relation**, R, is the **Union of sets** $R \cup R^{\sim}$, where $R^{\sim}$ is the **Relational inverse** of R. It is a symmetric relation.

Symmetric relation A **Homogeneous relation** for which x is related to y whenever y is related to x; in terms of **Sets**, R is a symmetric relation if and only if whenever $(x, y) \in R$ is true then $(y, x) \in R$ is also true. See also **Reflexive relation, Symmetric relation, Transitive relation** and **Equivalence relation**.

Target set See **Relation**.

Tautology A **Propositional form** which is always T no matter what propositions are substituted for the letters in the propositional form. See also **Contradiction**. A tautology P can be denoted by $\models P$.

Total function A **Function** in which the **Domain** is equal to the **Source set**.

Transitive closure The transitive closure, R^+, of a **Homogeneous relation**, R, on a set X is the 'smallest' **Subset** of the **Product of sets**, $X \times X$, which is a transitive relation and for which R is itself a subset. By 'smallest', it is meant that R^+ is a subset of any other transitive subset of $X \times X$ for which R is a subset.

Transitive relation A **Homogeneous relation** for which x is related to y and y is related to z only if x is also related to z; in terms of **Sets**, R is a transitive relation if and only if whenever $(x, y) \in R$ and $(y, z) \in R$ are true then it is also true that $(x, z) \in R$. See also **Reflexive relation**, **Symmetric relation**, **Transitive relation** and **Equivalence relation**.

Union of sets The union of sets A and B, denoted by $A \cup B$, is the set of elements in either A or B or possibly both.

Universal quantifier The symbol $\forall$, which binds the variable to which it is applied (making the variable a **Bound variable**); the **Proposition** $\forall x :\in A \bullet p(x)$ is T if and only if $p(x)$ is T for every member x of the set A.

APPENDIX D

Table of symbols

Logic	
Symbol	Name
Logical Connectives	
$\neg\, p$	Negation (NOT)
$p \vee q$	Disjunction (OR)
$p \wedge q$	Conjunction (AND)
$p \Rightarrow q$	Conditional
$p \Leftrightarrow q$	Biconditional
Quantifiers	
$\forall x :\in A \bullet p(x)$	Universal quantifier
$\exists x :\in A \bullet p(x)$	Existential quantifier
$\exists_1 x :\in A \bullet p(x)$	Unique existential quantifier
Metasymbols	
$\models$	Logical implication
$\equiv$	Equivalence

Special Sets		
Symbol	Set Name	Meaning
$\varnothing$	The empty set	$\{\}$
$\mathbb{Z}$	Integers	$\{\ldots, -2, -1, 0, 1, 2, \ldots\}$
$\mathbb{N}$	Natural numbers	$\{0, 1, 2, \ldots\}$
$\mathbb{N}_1$	Positive integers	$\{1, 2, 3, \ldots\}$

Sets	
Symbol	Meaning
$\{\ldots\}$	Denotes a set
$a \in A$	a is an element of A
$a \notin A$	a is not an element of A
$A \subseteq B$	A is a subset of B
$A \subset B$	A is a proper subset of B
$\#A$	Cardinality of A
$A \cup B$	Union of A and B
$A \cap B$	Intersection of A and B
$\bar{A}$	Absolute complement of A
$A \setminus B$	Complement of B relative to A
$A - B$	Relative complement – alternative notation
$A \triangle B$	Symmetric difference of A and B
$A \times B$	Cartesian product of A and B
$\mathbb{P}A$	Power set of A
$\mathbb{P}_1 A$	$\mathbb{P}A \setminus \varnothing$
$\bigcup A$	Generalized union of A
$\bigcap A$	Generalized intersection of A

Relations and Functions	
Symbol	Meaning
Relations	
$A \leftrightarrow B$	Set of relations from A to B
(a, b)	Ordered pair
$a \mapsto b$	Ordered pair (maplet notation)
$\operatorname{dom} R$	Domain of R
$\operatorname{ran} R$	Range of R
$R^\sim$	Relational inverse of R
$R \mathbin{\substack{\circ \\ 9}} S$	Relational composition
$R \oplus S$	(Relational) Override
R^+	Transitive closure of R
R^*	Transitive–reflexive closure of R
Functions	
$A \rightarrow B$	Set of functions: domain $= A$
$A \nrightarrow B$	Set of functions: domain $\subseteq A$
$A \rightarrowtail B$	Set of injective functions
$A \rightarrowtail\!\!\!\rightarrow B$	Set of bijective functions
$f \circ g$	Functional composition
$f \oplus g$	(Functional) Override
$[\![\ldots]\!]$	A multiset (bag)
$\langle \ldots \rangle$	A sequence

Index